TWILIGHT WARRIORS

TWILIGHT WARRIORS

Inside the World's Special Forces

MARTIN C. AROSTEGUI

St. Martin's Press
New York

ISBN 0-312-15234-5

First published in Great Britain by Bloomsbury

First U.S. Edition: February 1997

10   9   8   7   6   5   4   3   2   1

CONTENTS

1

GENESIS

'Get Hitler! . . . Preferably alive!' ordered General 'Wild Bill' Donovan, President Roosevelt's personally chosen director of the Allied Office of Strategic Services (OSS). He was at his Paris headquarters on the Champs-Elysées but could just as well have been back on the Río Grande, riding across into Mexico as a young cavalry officer in an expedition to capture Pancho Villa. It was March 1945 and intelligence from inside the collapsing Third Reich indicated that high-ranking Nazis, including the Führer himself, were going to take refuge in an impregnable redoubt area in the Austrian Alps, protected by a fanatical band of black-shirted SS troopers.

As Donovan spoke, Operation Iron Cross was being organized to infiltrate a guerrilla force into the mountains under the command of Captain Aaron Bank. The American 'Jedburgh' had operated in the vineyards of southern France, leading popular resistance forces against the Germans, months before the Normandy invasion. Dressed in the baggy attire of a French civil servant in order not to attract undue attention, Bank was visiting Allied POW camps to recruit turncoat German soldiers for his plan to penetrate the Austrian Alps. 'Tell Bank to get Hitler!' Donovan insisted, pounding his desk.

The Allied Jedburghs, so named after a twelfth-century clan of Scottish warriors, composed the special action branch of the OSS. They included a carefully selected cadre of American, British and French officers trained in the black arts of special operations at a requisitioned Elizabethan mansion in southern England. The manicured lawns of Milton Hall, which had once hosted elegant house parties and Sunday games of croquet, were now filled with dangerous young men stripped to the waist, practising unarmed combat, instinctive shooting and demolitions. After their crash course, the

Jedburghs would be parachuted behind enemy lines to assist organized resistance forces in Nazi-occupied countries.

In the Austrian Alps, however, there was no organized resistance to the Nazis and Captain Bank's team would be jumping into a void, setting up their own guerrilla movement. Working with another American officer and three sergeants, Bank had created a replica of Milton Hall at a French château in the Saint Germain forest, just outside Paris, where he was training a force of 170 German deserters in all aspects of clandestine warfare, including weapons handling, ambush and patrol tactics, demolitions, house clearance, sentry elimination, and survival. The special unit had been made operational in ten weeks.

In a meeting with Donovan's staff at the Paris OSS headquarters, Bank was shown aerial photographs of his proposed drop zone above Austria's Inn Valley at a snow-covered altitude of 4500 feet. He feared that some men and supply canisters might fall into a canyon at one end of the DZ. 'It is the most suitable,' was the curt reply when Bank asked if this was the only landing site available. 'OK, I'll chance it first with my pathfinder team,' he agreed.

'It was to be one of the most uniquely challenging special operations of World War II,' said Bank when I talked to him at his beachside home in San Clemente, California, where he lives with the attractive German woman he married back in his glamorous days as an OSS James Bond. 'The father of special forces,' as he is known, had just turned ninety and was writing a novel based on Operation Iron Cross. Although he was somewhat slow in speech and slightly hunched, his mind seemed to be as lucid as it was when he went over the plans at OSS headquarters overlooking the Arc de Triomphe.

'When told that my orders also included Hitler's capture, I went into a state of shocked disbelief mixed with elation. That night I lay awake for some time. Wild thoughts – world-shattering in scope – were racing through my mind. The implications of the expanded mission were starting to dawn on me. If we were fortunate enough to carry it off – get Hitler, alive rather than dead – it would be automatically the end of the war. The German general staff would surrender without a doubt. It would be the first time in my recollection of historical events that five guys would be responsible for ending a major war.'

While Aaron Bank prepared his special forces coup, the founder of the British Special Air Service (SAS), David Stirling, lay languishing in

Germany's maximum-security Colditz Castle prison. The unit he had created kept marauding behind German lines in their armed jeep patrols just as they had originally done in North Africa: sabotaging vital railways, attacking road convoys, blowing up bridges and raiding divisional headquarters. As the Allied armies began pushing into Germany, the SAS relished the task of rounding up high-ranking officers of the SS and Gestapo. There were some personal scores to settle. Captured members of the SAS had been tortured and killed by the Nazis following a personal directive issued by Hitler: 'Captured Special Forces troops must be handed over at once to the nearest Gestapo unit . . . these men are very dangerous and must be ruthlessly exterminated.' Stirling had managed to survive his two-year captivity by never letting the Germans know who he was.

Leading SAS operations in Europe was Colonel Paddy Mayne, who had served as Stirling's second in command in North Africa. A rugby player from Belfast with an evil temper and suspected by many to be a closet homosexual, Mayne was a man totally enamoured of action. Whenever there was a chance to attack the enemy, he would do so without a moment's hesitation. Ambushing a German supply convoy in northern France, Mayne drove his jeep straight into a German armoured car, getting so close that he could feel the sparks as his machine-gun bullets sprayed the vehicle. When return fire killed his gunner, he jumped out of the jeep, took the machine-gun and ran up to high ground, pouring more fire and lobbing hand-grenades down on to the Germans trying to take cover. It was with great difficulty that the other members of Mayne's patrol managed to tear him away from the fire-fight, in which they were desperately outnumbered.

Stirling had come up with the idea of the SAS while lying in a hospital bed in Cairo, suffering from a serious spine injury sustained in a parachuting accident. Things were not going well for the twenty-four-year-old lieutenant of the Scots Guards. A frustrated artist before the war, Stirling had mingled among café society in Paris, showing little interest in any established profession and often counting on his mother to cover his expensive hotel bills during frequent trips abroad. He had developed a passion for mountaineering and was climbing in the American Rockies, getting in shape for an expedition to scale Everest, when the war broke out.

Born to an aristocratic Scottish family, and blessed with a powerful six-foot frame, Stirling seemed ideally suited for his commission in the Brigade of Guards. But his performance was lacklustre at officers' school, where he often nodded off during tedious lectures on infantry tactics of the Great War. Going off to war with a 'fairly unfettered mind', he joined the newly formed Commando Corps, raiding German installations along the North African coast and the Greek islands. But it was not long before his original mind began taking a critical view of the way that commando operations were being conducted.

Conventionally, a force of 200 men would approach an enemy beach on board a large naval vessel, usually losing the element of surprise, which invariably led to casualties and damage before they got ashore. A large portion of the force got tied down holding the beachhead while the remainder went on to attack the airfield, radar station, supply dump or other rearguard facility. Stirling considered it to be an excessive expenditure of assets and manpower to achieve little more than pinprick effects upon the enemy. He wrote a memorandum to the commander of British forces in the Middle East, General Claude Auchinleck: 'Tiny raiding parties no bigger than four or five men would be far more elusive than a regular commando company of fifty or more. They can be dropped by parachute or infiltrated by jeep or small boats deep behind enemy lines to carry out reconnaissance and sabotage missions.'

It was a determined but gaunt and feverish Stirling who got up from his hospital bed one morning in July 1941, aided by a pair of crutches, to conduct his first raid for the SAS. There was no way that he was going to obtain an official pass to British Middle East Headquarters (MEHQ) by going through regular channels. The Adjutant General at MEHQ was his enemy – the instructor in whose lectures he used to fall asleep at officers' school. Stirling used to refer to him and his staff as 'layered fossilised shit'. A board of inquiry was also investigating Stirling's mounting gambling debts and licentious night life in Cairo. The honest opinion of a frequent gaming partner, Randolph Churchill, was that 'Stirling is too sensitive and doesn't really have the temperament to be a successful gambler'. He would often break into tears at the card table.

Several sentries were searching the MEHQ building for Stirling at the moment he burst unannounced into the office of the Deputy Chief of the General Staff. The commando had managed to slip into the compound through a tiny gap between the guardhouse and barbed-wire fence. A break-in had been reported and Stirling was going to be arrested. But General Ritchie immediately recognized the merits of the ideas which this fugitive lieutenant was rattling off to him. Not only could the low-cost means he was proposing to use to sabotage German supply lines and rearguard facilities be of strategic significance, but General Auchinleck was trying to find ways to tie down German manpower to stall the expected offensive into Egypt by General Rommel's Afrika Korps. There was no real native resistance against the Germans and Italians in North Africa and small units of men conducting guerrilla operations could provide the missing element to unbalance Rommel's juggernaut.

The Special Air Service which Stirling was given the go-ahead to form had little to do with the air. By the time they got into high gear, in 1942, SAS patrols were operating mainly from jeeps armed with twin Vickers anti-aircraft machine-guns. The only aerodynamic aspect of SAS operations were its targets: over 400 enemy aircraft destroyed by SAS raids on German and Italian airfields in North Africa – considerably more than the Royal Air Force had hit.

Although Stirling conceived of the SAS as an airborne unit, giving all of his initial recruits a parachute course which killed two of them, the first attempt to drop behind German lines was a total fiasco which almost depleted the embryonic regiment. Stirling had been advised to call off his first airdrop, which was planned as part of a general offensive Auchinleck was launching to relieve the besieged city of Tobruk in November 1941. The SAS had been tasked with raiding five German airfields in Libya but the weather report for 16 November was about the worst that could have been expected. In the early days of parachuting, winds above twenty knots were considered highly hazardous, causing units to scatter over such wide areas that they could not regroup and were forced to land violently. The forecast for Stirling's D-Day was for winds of thirty knots. But determined to establish his 'who dares wins' ethos from the start, Stirling insisted on going through with the mission.

One of the three Bristol Bombay aircraft carrying the raiding force was thrown off course by the high winds, was hit by enemy anti-aircraft fire, crash-landed and all twenty SAS men on board were taken prisoner. The teams in the other two planes managed to parachute out, but as predicted, suffered serious injuries when they hit the ground. Stirling himself lost consciousness from the shock of landing in a wind storm and awoke as he was being dragged at high speed by his parachute over gravel and rocks. It took him two hours to regroup his unit. One missing man was never found and about a third of the force were effectively out of action with sprained ankles and broken limbs. Most of their supplies, which had been dropped separately in air pallets, got lost. They could not find their weapons container, so were left armed with only their .38 revolvers and a few unfused bombs.

In the circumstances there was little they could do except head for their prearranged rendezvous with vehicles of the Long Range Desert Group (LRDG), a reconnaissance team made up of hardened New Zealanders, Australians and Rhodesians who became merged with the SAS. That meant walking over about fifty miles of open desert navigating on a compass with one water bottle each. They would have probably perished of thirst had the same forces of nature which had ruined the first SAS mission not saved them. The worst storm to hit North Africa in thirty years opened the skies in a spectacular downpour which turned dried-out wadis into torrential rivers. But if Stirling had not adjusted his bearings to take into account the distance which they had strayed from their intended landing point, he would have turned right, instead of left, at a desert track featured on their maps. One group which failed to do this fell into German hands.

It was the first SAS selection course. Of the sixty-two men who had set out on their first mission, only twenty-two made it to the RV. One was Pat Riley, a tall American from the Midwest who had got into the Coldstream Guards at the beginning of the war by lying about his nationality. He would become the SAS's first regimental sergeant major. Another was Jock Lewes, the demolitions expert who developed the 'Lewes bomb' which would wreak havoc in German installations during future SAS missions. Yet another was Johnny Cooper, who would remain with the SAS through the postwar campaigns of Malaya and Aden.

Another survivor of the first SAS mission was Paddy Mayne, who had been recruited out of a military prison in Egypt, where he awaited a court-martial for striking his commanding officer and chasing him out of the officers' mess with a bayonet. Stirling had managed to convince the aggrieved colonel to drop the charges by stressing that a unit like the SAS needed men with Mayne's temperament.

'He was naturally exuberant when being tested physically,' Stirling had observed about the vicious Ulsterman who was appointed second in command. In a practice run for their first mission, Mayne led a patrol on a ninety-mile march over the desert carrying rucksacks stuffed with stones and four pints of water each. They moved at night, lying up during the day in the flat desert with only pieces of hessian to protect them from the scorching sun as they lay motionless, repeatedly bitten by insects. One soldier started to lose his nerve and when his complaints grew too loud Mayne grabbed him with one hand and held him over a steep escarpment, saying, 'Any more from you and that's your lot!'

During the first SAS assault carried out some weeks after the disastrous airborne experiment, Mayne led a four-man team crawling beneath barbed wire into the German airfield of Tamit, in Libya, 400 miles behind enemy lines. After they had planted their bombs in twenty-four aircraft, Mayne decided to complete the task by bursting into the pilots' mess and spraying his Thompson sub-machine-gun until not a single German flyer was left moving. As he walked off the airfield amid the smoke and flames of the exploding aircraft, he noticed one Messerschmidt which had been overlooked. Having run out of bombs, he climbed into the cockpit and tore out the instrument panel with his bare hands.

Mayne returned to the same airfield a second time and blew up twenty-seven more aircraft. The jailbird was personally responsible for practically half of the 109 enemy planes destroyed in the first phase of SAS operations conducted by the twenty-two survivors of the failed parachute drop. Stories about the SAS's success soon spread like a new gospel throughout the Army and the regiment grew in size to several hundred men, to become a brigade. It took in other Allied recruits, including French Foreign Legionnaires, Greek resistance fighters and even knife-wielding cutthroats from Middle Eastern Mafia gangs. When Churchill visited Cairo he met David Stirling and took a great interest in the SAS. MEHQ officially acknowledged that the SAS had made a 'strategic contribution' during the North Africa campaign not only by denying Rommel control of the air but by making the railway supplying his front line virtually inoperable during the weeks preceding the decisive battle of El Alamein.

Later Stirling was captured by the Germans during an ill-advised attempt to drive straight across all of enemy-held North Africa for a symbolic link-up with the second Allied front being opened by newly landed American and Free French forces in Algeria.

'To find men of daring you must look to those who hold this world in poor account which includes both those who are outlaws of this world and in revolt against it and those whose thoughts and desires are fixed on the next,' is the observation of Charles Ogburn, an author and member of the 5307th Long Range Penetration Group, which operated behind Japanese lines in Burma. Among the constellation of secret armies and special units which formed in World War II, only to disappear afterwards, many were recruited out of the stockades in their respective military justice systems.

An American cameraman attached to Patton's 3rd Army distinctly remembers coming across a group of GIs training in southern England for a top-secret mission behind German lines. 'They all came from death row, were dirty, nasty and constantly fought among themselves. One was a Navaho Indian, another was black and others came from various ethnic backgrounds in urban cities. Some were dangerous psychotics charged with such offences as murder and rape who had been given the option of signing up for a suicide mission or facing execution.' The US Army has no official record of the unit's existence but the story gave rise to the legend of 'The Dirty Dozen'.

'The Army at times did arrange furloughs for various types of criminals who signed up for specially hazardous duties,' says Aaron Bank. Many jailbirds were to be found in the First Special Service Force, a mixed American and Canadian unit which became popularized as 'The Devil's Brigade' after being largely obliterated in a series of assaults on impregnable German positions during the Italian campaign. Its most famous mission was the taking of Monte La Difensa, a strategic 3120-foot mountain in southern Italy which regular Allied units had failed to capture in several attempts. 'The Black Devils', as they were called, because they were among the first to paint their faces with camouflage grease, climbed almost-sheer 1000-foot cliffs on a side of the mountain that no other Allied force had dared to assault and the Germans had not bothered to defend. They secured the mountain top in a six-hour fight after Allied generals had estimated that it would take six days.

Lieutenant Sam Wilson could have faced a court-martial for lying about his age under oath to get into the Army at the age of sixteen. But the Virginia farm boy proved so adept at light infantry skills that he was commissioned as an officer at eighteen and recruited into the OSS. He was assigned to train a group of Sicilian professional Mafia killers from New York who hated Mussolini and were going to be 'the unconventional component' of the Allied invasion of Italy. 'They were mean and scarred men and all of them had police records,' recalls Wilson. 'They knew how to use guns but had to be taught how to handle machine-guns and light mortars. They knew very well how to set up an ambush on a city street but needed to know how to do it in the woods.'

Wilson was among the first officers to be selected to go to Burma with the 5307th after answering a letter signed by President Roosevelt requesting volunteers for 'hazardous and dangerous missions'. He was immediately put in command of the intelligence and reconnaissance platoon and given authority to recruit his own men. He headed for the nearest stockade. 'Most were in for petty crimes like stealing rations, making bootleg whisky, shooting their weapons without orders and various charges of disorderly conduct.' But among them Wilson found 'brave men who held the rules of this world with some disdain and were comfortable with the calculated risk of death'. Another officer of the 5307th was an ordained Presbyterian minister who went through the chaplains' network to find men with 'strong spiritual beliefs who placed more emphasis in the world after death'.

At first the 5307th was going to come under the command of General Orde Wingate, a brilliant, highly religious and very neurotic recreation of Lawrence of Arabia who had led a camel expedition against Italian occupation forces in Ethiopia at the start of the war. A convinced Zionist, Wingate had once attempted suicide by cutting his own throat in a Cairo hotel because MEHQ would not follow his advice about forming a Jewish force for an offensive against Axis powers in the Middle East. Transferred to the South-East Asian theatre, Wingate organized locally recruited Chindit troops, commanded by British officers, to attack the Japanese in Burma, using elephants for supply trains. A last-minute change of plan, however, placed the 5307th under the command of the American General Joseph Stillwell, who was in charge of Nationalist Chinese forces with which the American detachment would be operating.

'I wouldn't want them in a garrison situation,' says Sam Wilson about his recruits. 'But if well led and trained, they could become very loyal and reliable, holding together better than most under the pressures of combat.' When the men of the 5307th were called on to form a guard of honour for Lord Louis Mountbatten, the British South-East Asian theatre commander, several of them refused. 'One just went AWOL for a couple of weeks in Calcutta, drinking it up and getting laid. No disciplinary action was taken against him because another group had gone off and hijacked a train for a tour of northern India.'

Under the code-name Operation Galahad, the 5307th's mission was to penetrate the thick jungle and high mountain passes of northern Burma to get behind Japanese lines and fall on the enemy positions from the rear, softening them up for a Chinese advance. 'The Marauders', as the unit became known, set a precedent as the first regiment-sized operation in history to be supplied entirely by air as they hacked their way through impregnable enemy-infested forests using horses and pack mules to bear their equipment. 'The Dakotas roaring in from Dijan in India provided almost our sole link with the known world, the main evidence that it was still there,' writes Charles Ogburn. 'We would stand about gazing up at them like the votaries of a cult that afforded infrequent but still impressive demonstrations of divine intervention.'

In contrast to the desert, where SAS patrols could get a clear view of any approaching enemy over vast distances, Marauders moving through narrow jungle tracks could constantly expect a hidden ambush. 'Uniforms were blotched with dark patches wet with the sweat of heat and the sweat of fear. Bend unfolded after bend and that was probably the worst part of the campaign – what was around the next bend. Hand raised . . . halt, stop, listen, on again . . . another halt . . . scouts went out, an almost unbearable sight, epitomizing the fate of man himself amid the emptiness and mystery that menace him. The platoon covering them as best they could from positions taken in the shadows waited for the horror of the machine-gun's clatter that would cut them down.' Sam Wilson survived a rifle barrage when he stumbled onto a hidden Japanese position less than 10 feet in front of him. A grenade which exploded as he hit the ground ripped through his knapsack as he managed to raise his rifle and shoot the enemy officer.

When they had to get off the tracks to avoid Japanese troop concentrations, the Marauders had to virtually dig tunnels through the solid jungle. 'We took in rotation the task of hacking a passage through the towering, tangled, resistant vegetation that buried hill-sides, valleys and ridges together and reduced us in scale to crawling animacules in its somnolent depths. You could hack a passage for days on end and nothing would happen because the tops of the growth were too interwoven for any of it to break loose. So the bamboo had to be cut twice, at ground level and at a height above the peaks of the mules' loads. The men struck savagely at the unyielding stems as if the vegetation were an enemy which had to be done to death until soaked with sweat, their arms heavy and gasping for breath, they would fall back to be replaced by others.' Wilson would have to get close enough to a Japanese artillery position to hear the breech-block clang and firing orders being called in order to pinpoint its location for an air strike. It was the only way to mount a reconnaissance in the cavernous foliage.

The Marauders were constantly bloodstained, not from enemy-inflicted wounds but from leech bites. Disease took a heavier toll on the men than did the Japanese. Dysentery became so pervasive that Wilson and others just cut out the seat of their trousers. Feverish and emaciated, they would plunge into savage combat. One company got surrounded on a hill, holding out for eleven days behind the cover of their decomposing dead mules and the piled bodies of Japanese soldiers who were being massacred as they attacked in waves.

'This morning we found that the Japs had grenaded one of our machine-gun positions and taken the machine-gun. They were firing it at us for an hour this morning. We know because they don't use tracer bullets and the gun they were firing had a lot of tracers. Some of the men on our south-west flank captured a Jap machine-gun this morning and fired all of the ammunition they had for it . . . the Japs broke through the perimeter in one place but were wiped out a few minutes later by two of our men throwing grenades. The hole in the perimeter fence was plugged but a chill swept through all of us as the story got around about how close the Japs had come to really getting in.'

By the time the 5307th was ordered to move on to its final objective, the Japanese airfield of Myitkyina, the Marauders were already overcome with disease and exhaustion. According to Ogburn, the first rumours of the impending ninety-mile death march through some of the highest peaks in northern Burma 'were like a tale of the supernatural in which from a mere suggestion of something abnormal a phenomenon of monstrous vitality and malignancy burgeons'. The men became dehumanized as they struggled through 6000-foot mountain passes which were greased with mud from monsoon rains. The mules would lose their footing and topple off the mountainsides, breaking their necks and stabbing themselves with bamboo. One hundred and fifty men succumbed to a typhus fever which struck suddenly, and had to be carried off on stretchers.

After the fighting in Myitkyina, the Marauders were exhausted. Most could not stand up for long without passing out, and they were all experiencing fever 'of unknown origin' with life-threatening temperatures of 105°F. Out of the original 3000 men, fewer than 200 remained in any fit condition. With the look of 'gaunt fanatic eyed bearded India Holy men', they lined up on the runway to be evacuated as the first American transports started to land on the captured airfield. Sam Wilson was defecating blood by the time he was carried by stretcher onto a plane for India. 'My last conscious thoughts were that I was going to die.' General Stillwell entered in his diary, 'Galahad is just shot.'

While they were decompressing from combat at their rest camp in India, the morale of the 5307th broke down completely. The criminal element seemed to take over as theft became rampant and the men ignored their officers, of whom few were left. Nights became drunken rampages, with the men, intoxicated with a local whisky laced with marijuana, firing their guns wildly into the air. There were screams from nurses being sexually assaulted while the mess became the permanent battleground of fist-fights and frenzied destruction. To call in the military police to maintain order threatened to result in a bloodbath and after their experience in Burma, the possibility of going back to the stockade paled into insignificance for most Marauders.

Wilson and other officers believe that among the reasons for the 5307th's breakdown was the complete lack of acknowledgement and recognition accorded the unit after the extreme hardship and suffering which they had bravely endured and fought through. Most of them had signed up for a three-month tour. The march through Burma had taken ten months. But there was not a single visit to their camp by any of the theatre commanders, nor any parade-ground ceremony or public occasion to honour them. No official identity or special status was ever conferred on the 5307th, which, deprived of an *esprit de corps*, just lost its cohesiveness.

It was meant to be that way. The Army had never considered the Marauders to be more than a 'provisional unit'. According to an official War Office memorandum written just after the Marauders had departed for India, the military high command were 'not enthusiastic about this type of organization and its special employment'. It was provided for one major mission which was expected to finish it more or less. In August 1944 the 5307th was officially disbanded.

When he first formed the SAS, David Stirling had been similarly denied permission by the military bureaucracy to create a regimental insignia for his unit. Being a detachment and not a regiment, the letter from MEHQ explained, the men would continue to wear the insignia from their original regiments. Stirling tore up the letter and got on with drawing up the flaming sword with the scrolled motto 'Who dares wins'. When it was finished, Stirling, the frustrated artist, had at last drawn his masterpiece. He ordered the device to be sewn on caps and painted on the vehicles, including his own convertible armed staff car. When a Chindit officer from Burma, Brigadier 'Mad Mike' Calvert, succeeded Stirling as SAS commander and presided over the unit's official disbandment in 1946, it was with the full honours of a Regimental parade.

One late-summer evening in Saint Germain, sentries kept guard around a large building housing the town hall, which looked onto a square intersected by four streets. The tables outside the café on the other side of the square were filled with people enjoying pre-dinner aperitifs and lively conversation.

'I don't see any of your crew except the enemy detail. When is the action going to take place,' asked the OSS colonel impatiently. 'Oh, in about five minutes,' Aaron Bank answered calmly.

A group of pedestrians passed by, couples walking arm and arm, all carrying bundles with wine, sausages, baguettes and other food for their dinner tables. At that moment about a dozen people rose from their tables, picked up their baskets and bags and started to walk towards the town hall, where women with enticingly heavy breasts were moving suggestively around the sentries. At the sound of a low whistle, German-made Schmeisser sub-machine-guns were pulled out of the bags and baskets. Flowers, baguettes and broken bottles of wine lay strewn all over the pavement as the mob suddenly rushed the building. The women's swirling skirts now exposed bulky hairy legs and tough, whiskered faces emerged from beneath the falling wigs.

Before they knew what was happening, the sentries were overpowered and the assault force divided into three groups. The main assault team burst into the town hall – enemy headquarters. The exterior security team took up positions outside the building while another group carrying a Panzerfaust anti-tank weapon headed into one of the side-streets.

It was a dress rehearsal for Operation Iron Cross – the mission to get Hitler, dead or alive – and the OSS officers observing it were duly impressed. One week later Bank was called back to OSS headquarters for a final briefing. After setting up operations in the Austrian Alps he would be liaising with the American 7th Army, which would soon be launching the offensive into Austria. Signals and codes were agreed on for Bank's German guerrilla force to identify themselves to the US forces. Bank was told to be prepared to go in a few days' time for his parachute insertion. The OSS had prepared falsified documents identifying him as Henri Marchand, a corporal in a German mountain brigade who had joined the Wehrmacht as a Vichy Frenchman from the island of Martinique.

German uniforms and weapons were among the supplies which Bank and his radio operator carried for their initial pathfinder drop, ahead of the main force, who would also go in disguised as German troops or civilians with forged Gestapo identity papers and ration books. But the Americans would not don their disguises until they were in Austria. They would parachute in their US Army uniforms with the standard .45 automatics strapped to their belts so that if caught on landing they would not be shot as spies.

Without informing the rest of his force that their mission had entered into its execution phase, Bank and his four-man pathfinder team were driven out to an OSS 'Joe house', as staging areas for secret missions were called. On arriving at a château located close to an airfield, they were issued with German money and gold coins. Bank was also given a big golden ring to bribe his way out of possible trouble. There were several other agents or 'Joes' waiting around to be flown out to various occupied countries. They all saw each other at mealtimes or walking around the grounds, but OSS officers supervising the Joe house made sure that strict silence was kept between them in order not to risk compromising any mission.

'The weather looks bad. The Alps are bogged down,' Bank was told the next morning by the Air Corps officer supervising the operations room at the airfield. They would have to wait until the following day for the flight. Bank was anxious. If his mission did not take off soon, the 7th Army would be moving into Austria before he did. He proposed risking a daytime jump. 'It's your neck,' was the reply from the operations room, which said they would consider it.

Bank, his communications sergeant and two of the Germans drank away the hours at the château's bar. Repeated phone calls to the air base elicited the same answer from the 8th Air Corps. The weather continued to be bad even for a daytime jump.

A week later Bank and his team were still at the château when they got news that the 7th Army had broken into Austria's Inn Valley and OSS central headquarters in London had cancelled their mission.

Hitler would commit suicide in his bunker six months later, on 30 April 1945, as the Soviet Union's Red Army overran Berlin. His body was found and incinerated by the Special Operations Unit of the Soviet Army, Smersh.

2

A DIFFERENT KIND

As terrorism and guerrilla warfare came to dominate the twilight struggle between East and West in what became known as the Cold War, President John F. Kennedy told the 1961 graduating class at the US military academy of West Point: 'This is another kind of war, new in its intensity, ancient in its origins – war by guerrillas, subversives, insurgents, assassins; war by ambush instead of combat; by infiltration instead of aggression; seeking victory by eroding and exhausting the enemy instead of engaging him. It requires a whole new kind of strategy, a wholly different kind of force and therefore a new and wholly different kind of military training.'

As long ago as 500 BC the Chinese General Sun Tzu wrote in *The Art of War* of the need for armies to develop units of what he called 'surviving spies' to conduct special operations behind enemy lines. 'Your surviving spy must be a man of keen intellect although in outward appearance a fool; of shabby exterior but with a will of iron. He must be active, robust, endowed with physical strength and courage, thoroughly accustomed to all sorts of dirty work; able to endure hunger and cold and to put up with shame and ignominy.'

Many of World War II's surviving spies were soon back in action reorganizing special forces units which for the first time were being assigned permanent roles in the order of battle. Veterans of David Stirling's SAS were recalled to fight a protracted guerrilla struggle against a Maoist uprising in Malaya. A French Jedburgh officer, Lieutenant Colonel Aussaresses, was ordered to form the 11ème Bataillon de Choc to deal with growing communist insurgencies in France's colonies. Aaron Bank began building the US Army's Special Forces, to fight behind the lines in Soviet-controlled Eastern Europe.

A core of World War II SAS officers, combined with volunteer detachments raised in Rhodesia, New Zealand and Australia, renegades from other British units and even a group of deserters from the French Foreign Legion, quickly adapted to Malaya's jungle environment. Operating in small bands of rarely more than four, they hunted down an elusive enemy throughout the country's thick forests. Their strategy consisted of denying the communists access to food or to populated areas while winning the 'hearts and minds' of local inhabitants with protection, medical care and other assistance. It is estimated that 1500 man-hours were expended on observation and reconnaissance for every contact made with the 'CTs' (communist terrorists) during a guerrilla campaign which lasted nearly a decade.

Training mainly on the job, the SAS learned the new rules of jungle warfare with the assistance of local tribesmen; for example, how to read footmarks along narrow jungle tracks, distinguishing the splayed toe prints of the barefoot native from the cramped ones of men used to wearing shoes. The latter indicated the passage of CTs since most of Malaya's communist guerrillas came from urban areas. During one patrol an SAS sergeant followed a set of cramped toe prints for several days, at times even searching through piles of stinking elephant dung. They eventually led him to a CT hideout, where he surprised four terrorists taking cover from a monsoon rainstorm and shot them dead at point-blank range with his repeater shotgun.

While parachuting into canopy jungle where trees grow to up to 200 feet, the SAS were led to experiment with tree jumping. Since parachutes inevitably got caught up in high branches, a man would have to attach a rope to the nearest tree and abseil to the ground. There were serious accidents, as when Sergeant Johnny Cooper, a survivor of Stirling's first parachuting fiasco in North Africa, found himself hanging 100 feet above a bamboo thicket, with a broken arm and his harness practically strangling him. With the help of a medic who climbed up and attached Cooper's rope to the closest tree, the SAS veteran was finally cut loose from his chute and allowed to drop for over 100 feet until the rope yanked him up. In the process he swung into a tree trunk, which smashed his spine, but he survived.

By 1958 the SAS was leading the manhunt for the last surviving CT leader, Ah Hoi, known as the Baby Killer for his public disembowelment of a pregnant woman. Having traced his hideout to an area of mangrove swamp measuring some eighteen miles by ten miles, a troop from B Squadron commanded by Captain Peter de la Billière had to wade through waist-high mud for days. Weighed down by 40lb rucksacks as huge leeches clung to all parts of their bodies, they silently crept up on one abandoned camp-site after another, hoping to surprise Ah Hoi.

But the noise from a helicopter which had been called in to evacuate a patrol member who had broken his back in the tree-jumping insertion had betrayed their presence and Ah Hoi was managing to stay one step ahead of them. As the patrol entered its eighth day the men were becoming worn down and running critically low on food and medical supplies. It was not possible to call for an air resupply as its arrival would further alert their prey.

With the success of the operation hanging in the balance, an SAS patrol approaching from another direction sighted some of Ah Hoi's men drinking water from an open river. Concealed by a floating log, two SAS men floated downstream to within fifty yards of the terrorists and opened fire. After killing two of the men, they discovered Ah Hoi's fresh camp-site, and although he narrowly managed to elude them, they had reinforcements airlifted in and a cordon was thrown up around the area. Hopelessly boxed in, Ah Hoi gave himself up after a few days.

After this there was never again a question of the SAS being disbanded. Instead, the Regiment was expanded to four squadrons, including the newly attached D Squadron, composed of volunteers from the Royal Parachute Regiment, and G Squadron, drawn from the Division of Guards. A group of black Fijians recruited through New Zealand were kept on as they had proved to be highly dedicated and resilient soldiers. Also, a rigid selection process was instituted to filter into the Regiment the type of recruits most closely resembling Sun Tzu's 'surviving spy'.

Captain Recolle of the 11ème Bataillon de Choc was having a heated argument with Ahmed Bellounis, a Berber chieftain being supported by France to counter the spreading influence of the Marxist Front de Libération Nationale (FLN) in Algeria. Despite the extensive training, money and arms which Recolle's special forces team had supplied to Bellounis, his native army had failed to prevent the FLN's takeover of the nearby village of Melouza, whose population had been massacred. As the recriminations grew ever more intense, the Berber chief picked up the pistol lying on his desk and shot his French military adviser through the heart. The French appeared to be losing the battle for hearts and minds in Algeria as well as elsewhere.

In Indochina, Ho Chi Minh's Viet Minh guerrillas, substantially backed by China and the Soviet Union, had besieged the French fortress of Dien Bien Phu. Up against a well-armed 'People's Army' of tens of thousands, supported by heavy artillery and the majority of the population, there was little that the battalion could do as the last colonial stronghold was overrun, except rescue a handful of French soldiers. Raids against Viet Minh supply convoys and some divisional headquarters by 11ème Choc and other paracommando units raised among the traditionally elite Foreign Legion were having little effect on a vast enemy. Vietnam was no Malaya.

In Algeria the Foreign Legion's 1 and 2 REP (Regiments Etrangers de Parachutistes) had managed to effectively seal off the country's borders against battalion-sized incursions by the FLN. But although France retained control of the main centres of population, strategic ports, and oil and gas fields, the FLN countered with terrorist campaigns. Undercover teams drawn from parachute units and pro-French *pieds noirs* of the Organisation Algérie Secrète fought back with a dirty war in the cities. However, by April 1961 France's President De Gaulle conceded Algerian independence to the FLN. Considering this the worst form of humiliation and treason, French elite special units turned against their own government.

Four generals installed in the colonial governor's palace in Algiers called on the army to topple De Gaulle. They were followed by the 1 REP, the Bataillon de Commandos Aéroportés and others. Near Paris, several officers of the 11ème Choc were arrested on charges of conspiring to extend the coup into France. 'If our officers thought it was a good idea then it was a good idea,' is how Tony Hunter Choate, a British volunteer in the 1 REP who later joined the SAS, describes the situation when his unit occupied the broadcasting centre, police stations and other key buildings in Algiers. 'We felt like we could hold out for ever.'

But the commander of the Legion's other parachute unit, the 2 REP, called Petit Pierre because of his diminutive height, contributed to the coup's eventual collapse. Realizing what was happening on the morning of the coup, he managed to order his regiment back as their trucks crashed through loyalist roadblocks on the highway to Algiers. While the 1 REP and other units were disbanded after the failed putsch, the 2 REP was retained, to develop into one of the most elite special operations units in the French Army.

* * *

Far from being limited to counter-insurgency, US special operations during the Cold War were altogether more ambitious. Since the late 1940s the USA's newly formed Central Intelligence Agency (CIA) had been engaged in efforts to spark insurrection inside Soviet-occupied countries. Among the special operatives assigned to such missions was Sam Wilson, a teenage World War II Marauder who had narrowly escaped death from typhus in Burma. Later, still very weak physically, he failed his medical examination for entry into West Point, but was recruited by the CIA instead. After some specialized language training, Wilson was sent to Germany to recruit and train Russians seeking to escape repatriation to the Soviet Union. (Considering their loyalty suspect, Stalin was sending many of his troops returning from Germany to work camps in Siberia.)

'We were forming an organization,' explains Wilson, 'to do unto the new Soviet rulers what Lenin had done to the Czar.' By 1949 the CIA was parachuting its Russian 'contras' into the Soviet Union with intricately forged identity papers, to organize a guerrilla movement against Stalin. At least one American officer of Slavic origin with a perfect knowledge of Russian went in with the secret army.

But operating in Stalin's communist police state proved immensely difficult if not altogether impossible. Organized internal opposition was almost non-existent and the CIA's guerrilla cells soon found themselves in a vacuum. 'We had very limited results,' says Wilson. 'Some carried out their missions, meeting with minor success.' Some armed guerrilla operations got underway in the Ukraine, and in Bulgaria and other East European countries. 'But many more agents were captured, tortured and executed. A few were even turned against us.'

The failure of the CIA to start a revolution in the Soviet Union spurred the US Army to create its own special forces unit. Aaron Bank had long proposed the idea, devising a military training programme to cover all aspects of unconventional warfare: 'Organization, tactics and logistics of resistance movements; espionage work and sabotage against railroads, highways, maritime facilities, airports, telecommunications, electric power lines, radar networks, specialized demolitions, the encyphering and decyphering of codes and clandestine communications.' Being of Russian immigrant parents himself, Bank went out of his way to recruit East Europeans. Among them was Larry Alan Thorne, a decorated Finnish soldier and arctic warfare expert who had joined Germany's SS to fight against the Russians in

World War II. Before becoming a sergeant in the new special forces, Thorne had already operated for the CIA.

Bank's 10th Special Forces Group 'would be used in a stay behind role' according to Sam Wilson, who was reassigned to the Army, developing plans for 10 SFG's deployment in Europe at the time of the 1953 anti-communist uprising in East Berlin. 'If the Soviets overran Berlin small teams would remain hidden in place to pop back up after the tanks had rolled over to organize the nucleus of resistance cells which would quickly expand guerrilla activities.' The idea was even proposed of prepositioning teams in 'freeze dried boxes' inside Soviet territory to eliminate communications and supply problems until they were thawed back to life at the outbreak of hostilities. The environment envisaged for future special operations was mainly urban, and action teams were mostly to be disguised as local civilians. The first US Special Forces shoulder patch depicted the Trojan horse.

The SAS was assigned a similar role in NATO plans, including the 'kidnapping of important enemy personalities and attacks on important industrial targets invulnerable from the air'. Special swimmer teams would also be inserted to sabotage Baltic sea ports in East Germany, Poland and Russia. For such missions the Royal Marines had the Special Boat Section (SBS), a handpicked detachment of sixty-three officers and NCOs. They had already engaged in night-time sabotage raids and reconnaissance along the North Korean coast, operating jointly with the US Navy's Underwater Demolitions Teams, forerunners of the Sea-Air-Land Navy Commandos, or SEALs, who were officially commissioned in 1961.

Another CIA failure, this time in Cuba, just ninety miles from American shores, prompted President Kennedy to further expand the role of Army and Navy special forces. In April 1961 some 1500 Cuban exiles trained and financed by the Agency were all killed or captured as they landed at the Bay of Pigs to topple the revolutionary communist regime of Fidel Castro. 'The military would have done a better reconnaissance of the beaches, inserted our own teams ahead of time to destroy Castro's air force on the ground and prevent his forces from concentrating. We would have trained the Cuban exiles better, would have planned strategically and tactically better,' claims Aaron Bank. 'Cuba was a sadly lost opportunity.'

'I am directing the Secretary of Defense to expand rapidly and substantially, in cooperation with our allies, the orientation of existing forces to conduct non-nuclear war, paramilitary operations and sublimated or unconventional warfare,' Kennedy told Congress in a speech immediately following the Cuban fiasco. He signed a classified directive removing the responsibility for paramilitary operations from the CIA and placing it with the military, while asking for the resignation of long-time CIA director Allen Dulles.

With a much-expanded budget, thousands of men were recruited among airborne forces, navy divers and other elite units, to form several Special Forces Groups targeted on various areas of the world. Language instruction ranged widely, from Spanish to Swahili. Sam Wilson, who was appointed director of training at the John F. Kennedy Special Warfare Center in Fort Bragg, North Carolina, recalls how in those days special forces took on an almost messianic character. Kennedy authorized them to wear the Green Beret as 'a symbol of excellence, a badge of courage and a mark of distinction in the fight for freedom'.

The President made inspection visits to Fort Bragg to see his supermen break piles of bricks with karate chops, climb 500-foot cliffs like spiders, do fifty left-handed push-ups and shoot out eye-sized targets at sixty feet. Here were the soldiers for Kennedy's 'New Frontier'; they were the incarnation of American strength and idealism; John Wayne with a global mission. The crossed arrows of the legendary 7th Cavalry's Indian scouts were adopted for the unit's new patch, superimposed on a dagger with the enscrolled Latin motto 'De Oppresso Liber'.

As the US military involvement in Vietnam grew and the first Green Beret 'mobile training teams' were sent to South-East Asia, an exchange programme was instituted with the SAS, which already had over a decade of success in counter-insurgency operations. SAS commander John Woodhouse often came to Fort Bragg, bringing along his top jungle-warfare expert, Sergeant Lofty Wiseman, who stayed on as an instructor. After Malaya, Wiseman had served in Borneo, where the SAS was again engaged in deep jungle patrols along a 700-mile border of thick tropical rainforest, fighting off armed incursions from neighbouring Indonesia.

A wild-eyed native of London's East End, Lofty was among the first to lead 'claret' operations – secret cross-border raids by tiny patrols, sometimes of no more than three men, where it was customary to 'shoot and scoot'. If a man was wounded in an encounter with Indonesian troops it was understood that he would be left behind. The SAS could count on no type of support from any home base. In one Borneo episode a sergeant, wounded in the legs as he covered the escape of the rest of his patrol, dragged himself for several miles to the nearest border. In another operation Lofty had to kidnap and bring back across the border a treacherous chieftain, or *pughulu*, of a head-hunting tribe.

'Adaptability' was the main quality Wiseman tried to instil in his American trainees, who he felt to be too encumbered by preconceived plans and field manuals. He tried to school them in 'the basics'. 'You don't need all that kit they give you – only as much food and ammo as you can pack in your bergen,' he would say in his high voice, which was also permanently hoarse from all the malarial coughs he had contracted.

Being among only 8 out of 125 who passed the first official SAS selection course in which 'weights got heavier and distances got longer until you were up against fifty-five miles carrying 55lb bergens', Lofty felt that 'Americans try to take the hard world out of soldiering'. Pushing a group of rain-soaked Green Berets through a survival course in Florida's Everglades, he would tell them: 'Use bad weather to your advantage. That's when the enemy has his head down so that's when you move.' He also reminded them that in the SAS it was necessary to sign a document saying that 'I agree to carry out arduous duties with no recognition, no reward, no promotion and no medals'. American special forces officers were also sent on attachment to the SAS base on the edge of Hereford, set among the hills and cow pastures of England's border with Wales. Captain Charles Beckwith discovered that 'all there was for me to do there was learn. I had very little to teach the SAS'.

In October 1962 U-2 spy planes confirmed reports that the Soviet Union was secretly basing medium-range nuclear missiles in Cuba, where they could threaten the entire continental United States. As Kennedy ordered a naval blockade around Cuba, the US Army's newly formed the 5th Special Forces Group started rehearsing direct assaults on the missile emplacements in mock-ups of the installations built at a training base in Panama. Limited commando raids on Cuba were seriously considered as a possible way to avoid all-out war with the Soviets at a point during the crisis when Moscow seemed prepared to take out the missiles but Castro was refusing to agree.

Thrown into action just some months after their creation, Navy SEALs arrived in submarines in Cuban waters. One squad surfacing into the night's blackness immediately saw three Soviet-built

Komar-class gunboats heading towards them. They swam away from the submarine as fast as possible since it had probably been picked up by Cuban sonar. With their recently perfected underwater rebreathable aqualungs, the SEALs could swim below the surface undetected as they were practically deafened by the sharp buzz of the patrols' engines racing right over them. After swimming for several miles, they reached Cuba that night.

The SEALs could see apartment lights shining as they climbed the beach fronting downtown Havana. Spotting a Cuban police patrol, they unsheathed their USN Mk-2 combat knives for a silent sentry elimination. Another team of eight swimmers stealthily made their way underwater into the Cuban naval harbour of San Mariel. After studying the hulls of some of the ships, they prepared to sink them with limpet mines. Plans were developed for SEALs to proceed inland after conducting operations along the coast and link up with Army special forces being parachuted in. However, the Soviet missiles were withdrawn and finally no offensive action was taken in Cuba.

It would be in neither Cuba nor Eastern Europe that America's new special forces would be tested, but in the killing fields of South-East Asia, where the French had already been defeated.

President Kennedy was assassinated a year after the Cuban crisis by a mysterious ex-US intelligence employee called Lee Harvey Oswald, who had defected and then redefected from the Soviet Union. The Green Berets formed the guard of honour at Kennedy's state funeral.

3

KNIGHTS OF THE APOCALYPSE

Struggling against the jungle sun, halfway up the summit of Nui Ba Den, to storm an impregnable Viet Cong fortress, Captain Bo Gritz paused in his sweat-drenched 'tiger suit' and turned to look upon fifty hearts and minds who would no longer follow. They were native Cambodians of his specially trained 'Cam Tu', Vietnamese for suicide: teams who had been personally selected by the Green Beret officer for this impossible 'one-way' mission. To qualify, Gritz had required each of them to hold a fused hand-grenade for thirty seconds, without telling them that the explosive charge had been removed.

Nui Ba Den, the Mountain of the Black Virgin, derived its name from an ancient Buddhist superstition about a young girl who had supposedly perished there in search of her lost lover. The black granite rock, mainly covered with thick forest, rose like a colossus out of the surrounding jungle with a commanding a view for over fifty miles. Men who were among the toughest mercenaries in Indochina now prepared to die there as the black virgin had done, sitting in a near state of catatonia, biting into small Buddha statues which each carried strung around his neck to assure a smoother passage to Nirvana. They had resigned themselves to death from thirst. Gritz had failed to take the precaution of packing in extra water for their mission and there was none to be found on his chosen approach to the mountain top. They had all squeezed the last drops out of their canteens that morning.

Gritz stomped down the trail, throwing down his carbine. 'Tell Hai Muoi that he and all his F team can lay right there. In a day or two the wild hogs will have cleaned their bones so they can bleach in the damned sun. The buzzards will pick them clean and after the war they can all come and see, like Custer's last stand, where F22 gave up

the ghost.' But the Cambodians remained unmoved, just lying there
leering at him with their Buddhas clenched between their teeth. Gritz
seriously thought that he might get shot and reminded them that they
had signed an oath in blood and sworn on the bones of their ancestors
that they would follow his orders to the grave.

The other five Green Berets in Gritz's A team staggered up to him,
trying to describe the men's condition: '. . . passing out . . . heat
cramps . . . heat strokes . . .', but otherwise offering little in the way
of encouraging advice. 'The burden of command bore down on me
with more intensity than the sun or the sack on my back,' recalls
Gritz, who now turned towards the heavens and prayed.

It was 1965 and there were already 15,000 American troops
defending the Republic of South Vietnam against Ho Chi Minh's
communist North and his Viet Cong guerrilla allies. The special
warfare advocates like Lansdale and Yarborough whom Kennedy
had cultivated had been cast aside by the conventional military top
brass, who were advising President Johnson to escalate the war to
unprecedented levels. The US Army's 1st Infantry Division, the 'Big
Red One', stood positioned in the jungle below Nui Ba Den, having
repeatedly attempted to storm the well-entrenched Viet Cong positions commanded by Major Mong, one of the most feared communist
guerrilla leaders. Just as repeatedly, the regular American troops had
been repulsed. It was Gritz who had personally volunteered to lead an
assault with his Cam Tu teams through the back door of the fortress
carved out of the huge boulders and caves on the slopes facing east
and which had solidly resisted air bombardment. It was considered a
suicide mission and Gritz was being allowed to try it.

Turning back again to look at his men after his silent heart-to-heart
with the Almighty, Gritz felt a calm confidence come over him as
David, the senior among the Cambodians, walked up to him and said,
'It has been decided: we will stay here.'

'Dave,' Gritz calmly replied, 'you have three choices. You can stack
everything you have in a big pile since everything you have belongs to
Uncle Sam. You are then free to do anything you like. Or you can
keep everything and follow me and we will press on together as
friends, or you can keep all the gear and become the enemy when the
American team moves out. The choice is yours. You've got fifteen
minutes.'

Gritz prepared to carry out his threat. He gestured to his radio man
as he picked up his PRC-25. 'Smokey, this is Swamp Fox, over,' he

was calling up to a half squadron of A1-E fighters providing his air cover.

'Hello Swamp Fox, this is Smokey. I've got four fast movers up here from Bien Hoa with a load of red hots for Charlie. Where do you want them?'

'Smokey, pick up my signal mirror and make the first pass a dry one right over my position with your tail between your legs and kindly light the afterburners on pull-up. Make the run parallel to the mountain. I've got a small leadership problem down here and if I don't get it settled on the first run you can make a second pass twenty millimetres down the flanks. If we've got no joy by then you can dump the nape right here.'

As the silvery shapes of the four fighters became visible against the sky, there was a thunderous roar which shook the ground and sucked the breath out of one's lungs. In the next split second, only the flames from their engines could be seen becoming small dots as they disappeared into the high clouds. An explosion followed as the jets went into overboost, making the rocks vibrate throughout the mountain like an avalanche.

Gritz's eyes met Dave's, who looked at him as if asking for compassion. No words were exchanged. Smokey came back on the radio. Gritz repeated the instructions. 'Roger, Swamp Fox. Lead in hot this time.'

The crack of exploding 20mm shells raining towards the Cambodians preceded the roar of the jet engines on their second run as their GAU-8 Gatling guns put down 6000 rounds a minute. Everyone on the mountainside was showered with the falling brass cartridges and pandemonium broke out. The Cambodians burst into screaming fits and a frenzy of activity. Then order emerged from chaos, and they were soon neatly lined up in two single files, prepared to follow Gritz and the other five Green Berets up the mountain.

Gritz believes that it was the answer to his prayer when a shriek went up from one of the Cambodians, who had discovered an underground stream of fresh clear water seeping through the rocks. 'We all filled up our canteens and morale soared.'

No sooner had Gritz and his Cam Tus reached above the cloud cover and begun their approach on Mong's rear than they were pinned down across a steep ravine by automatic fire and rifle grenades. He signalled the A1-Es for a napalm strike on the VC and the four jets roared in, dropping the jellied gasoline, which just burned uselessly at the top of the 150-foot tree canopy.

The only way across the ravine was over a flimsy rope suspension bridge. With his men positioned to cover him, Gritz was preparing to race across when the ground shook with an exploding 250lb bomb dropped by one of the planes. Dirt, foliage and a human leg rained down on them. 'Sandal!' said one of the men checking the footwear on the leg, indicating that a nearby VC position had taken a direct hit.

Gritz now made for the bridge, running through a barrage of automatic fire which forced him to land face down on the narrow foot planks swaying from side to side. As he inched up his head, he could see the shape of an unexploded 100lb bomb half buried in the earth in front of him. Firing off the thirty-round clip of his M-2 carbine, he leaped onto to the edge of the ravine and took cover behind the bomb. He reloaded and fired off another clip, covering the rest of his men following him across the bridge, who fanned out around him. As they all advanced, the enemy's automatic fire died down, being replaced by mortar shells which fell inaccurately in the thick jungle. Shrieks suddenly erupted as an explosion ripped through a group of Cambodians at the front. One Cam Tu pointed towards the ground in front, gasping out, 'Mine.' They had hit a minefield: Major Mong's last line of defence against a rear intrusion into his fortress.

A pass had to be cleared before Mong could rally his troops for a counter-attack. The Americans would have to do it: the Cambodians were scared of mines. Gritz crawled forward with another Green Beret, Sergeant McNeal, pushing aside the smouldering torso of one of their dead, which swamped them with the odour of blood and ammonia. What seemed like gallons of blood was splattered on nearby trees. Gritz could make out a maze of trip-wires attached and leading to fuses on hidden bottle mines, home-made fragmentation mines and more sophisticated Chinese DH-10 disc mines.

The Cambodians sat in the protection of the jungle eating rice as the other Green Berets laid on covering fire for the mine-clearing team. In less than thirty minutes they had defused enough mines to clear a passage, and at the end of the trail Gritz saw the actual back door of the fortress: a bamboo gate with attached disc mines. Firing wires disappeared into the rocks and the natural cover, the latter woven to form an almost solid wall. He told McNeal to go back and start moving the rest of the force through the pass and bring up two M-72 LAWs (Light Anti-tank Weapons) to blast the mined gate.

As the sergeant turned to go, a bullet whizzed inches above Gritz's head and the next thing he knew, McNeal was gasping next to him with a geyser of blood spurting from somewhere close to his groin. Screaming out his orders, Gritz pulled the sweat band off his head, wrapped it around McNeal's crutch and applied a tourniquet to his right leg. By pushing a small rock between the strap and the flesh, he stemmed the gushing blood. Gritz yelled again for his radio to call in a medevac.

'It's not my balls, is it, Bo? It's not my balls,' McNeal asked in a quivering voice. 'No, Mac, it's not your balls,' Gritz replied, quite certain that the wound was just above the leg.

The rest of the force was moving forward. Two explosions signalled that the gate was blown open. The medic rushed up to look after McNeal, who was using up his failing energies to feel around his crutch, making sure that his testicles were still intact.

Gritz now moved into the battle as his men assaulted Mong's invincible fortress, letting out crazed victory shouts of 'Tay Yo' while pouring automatic fire and grenades into the enemy emplacements, which were wide open from the rear. The VC had fashioned fighting positions out of the boulders. Interlocking fields of fire had been carefully cleared which sighted down on open approaches to the mountain. Terraced rice paddies were laced with mines and traps. The high primary jungle provided an umbrella against napalm while the thick secondary vegetation absorbed cluster bombs and 20mm cannon fire. But Mong had never expected to be attacked by a group of fifty men capable of sneaking down from the peaks above him.

Hardly needing any further direction from the Green Berets, the Cambodians pressed forward their savage assault, doing what they did best – killing VC. They were having little trouble clearing the enemy positions as the shocked enemy scattered and ran from their bunkers. Then Gritz heard to the back of him the sound of a bolt going forward, followed by the frantic manipulation of the bolt operating handle. Spinning around, he looked straight into the slant eyes of an Oriental in black pyjamas and a green floppy hat holding a Chinese K-50 7.62mm machine-gun. He hosed down the VC with the half magazine left in his carbine, watching the blood-splattered body fall back into the bamboo-covered hole. If the machine-gun hadn't jammed, Gritz would have been dead. 'God was with me that day.'

'Gritz, are you all right? I knew you were a piss-cutter. Only dipshits would let a little thing like a head wound take them out. Get that needle out your arm and let's get out of here. We've just been committed to another mission up in II Corps.' It was Gritz's new commander, Major Charles Beckwith of Project Delta, a bear of a man with a thick Georgia drawl and a mean look in his eye, who could push generals around and was one of the first American special forces officers to train with the SAS.

Gritz hadn't been as lucky in his last mission and a bullet had grazed his skull as he tried to take some North Vietnamese prisoners near the Cambodian border and now Chargin' Charlie had erupted into his hospital room to ruin his convalescence. Ignoring the head nurse's protests, Beckwith pulled the intravenous needle out of Gritz's arm and told him to get into his uniform, adding: 'Let's clear this place. I've got a jeep outside with the motor runnin'.'

When Beckwith had first arrived to assume command of the special reconnaissance group, Detachment Delta, he had found their beach-side camp empty. On driving into the nearby town of Nha Trang he was greeted by all of the officers, waving at him from the windows of a run-down hotel they called the Jockey Club, with naked Vietnamese girls in their arms. 'What kind of a goddamn war are we fighting here?' exclaimed Charlie, as he stood up in his jeep, gazing around in amazement.

If the sixties was the age of tolerance and free love, Beckwith hadn't heard about it and he proceeded to fire all but seven of the forty men in Detachment Delta, casting them off to less pleasant assignments, like manning an observation post on Nui Ba Den. The first to go was the senior sergeant, who didn't mind admitting that he was running the Jockey Club at a profit. Delta's new commander now took the opportunity to create his own special forces unit, sending flyers throughout the entire US Army in Vietnam, saying 'WANTED: volunteers for Project DELTA. Will guarantee you a medal, a body bag, or both.'

To his surprise more volunteers turned up than he could handle: Green Berets, airborne rangers – now starting to come into Vietnam – and some regular soldiers looking for excitement. He got complaints from many unit commanders saying that he was drawing away their best men and that's exactly what he wanted to hear. 'I'm ruthless,' Charlie relishes saying.

He ran all of his volunteers through an SAS-type selection course on an offshore island. 'It was a tough sonofabitch. When we got them there, we made them move from one point to another, fast. No high trees but tough scrub country, very difficult terrain, and they had the heat leaning on them because there was no canopy keeping the sun out. Most collapsed from exhaustion – that's what I wanted. I wanted them to quit.'

Beckwith's strongest, most deeply felt and innermost desire had been to create his own elite unit ever since spending his year on attachment to the SAS. When he saw how troopers who missed their rendezvous during navigational training were dunked in freezing cold water and forced to remain in their soaking uniforms for the rest of the exercise, he thought, God, this is what we ought to have in our army.

Being the kind of officer who never stopped polishing his boots, Beckwith was at first taken aback by the informal ways of the SAS, where soldiers were allowed to wear their hair long, keep their barracks messy, and other normal army rules were not enforced. But gradually he came around to see the value of the close camaraderie it engendered. When the officers were invited over to the NCOs' mess on a Saturday night to stay up until the daylight hours talking, exchanging ideas and drinking beer, the regiment seemed more like a club and he liked its cosiness. Receiving the sand-coloured beret with the flaming sword after completing a three-day navigation course in which his feet got so badly blistered that they were bleeding, was an event in Beckwith's life. 'You had to earn this one,' he said. 'My green beret was just handed to me.' He marvelled at how SAS officers had their berets custom-made.

The SAS took Beckwith on an exercise with the French 11ème Bataillon Parachutiste de Choc in Corsica, where they practised storming into a sixteenth-century fortress and then escape and evasion through the island's rugged hills. He was then taken on a jungle training operation in Malaya, where they practised immediate action drills firing live ammunition. 'It was great, I had never done this in the States.' He admired how his troop instinctively handled every contingency. If rafts had to be built to cross a river, half the men started gathering bamboo while the rest gathered rattan with which to tie the sticks together. 'No orders were given, no tasks were assigned, everybody just automatically did their part.'

'Charlie just couldn't believe how we laughed and joked when we were up to our heads in swamp water,' recalls Lofty Wiseman. 'It's the only thing you can do sometimes to keep from crying. He soon learned to laugh with us as well.' When Beckwith decided to bathe in a jungle stream, the SAS men tried to stop him. 'What you want is to build a thick crust all over you so that the bloody mosquitoes can't bite through.' Beckwith went ahead and bathed anyway. He ended his jungle exercise with a severe case of leptospirosis, a potentially fatal disease caught from the urine of infected animals in jungle rivers. For Beckwith, it meant several weeks of hospitalization with penicillin injections every three hours.

He came back to Fort Bragg full of ideas about how to apply the 'SAS concept' to create a similar unit in the US Army. Beckwith believed that US special forces had expanded too quickly under Kennedy and a lot of unqualified personnel were being accepted for the sake of filling spare troop capacity. 'I'd rather have ten guys who can paddle than a hundred shitheads.' Bo Gritz, for example, had come straight into the Green Berets on joining the army, without first serving in any other regular unit, but 'this would not be acceptable in the SAS.'

But nobody seemed interested in listening to Beckwith and he was told to go back to school and finish his degree. Even General Yarborough, the normally open-minded and gentlemanly director of Fort Bragg's John F. Kennedy Special Warfare Center, told Beckwith that he had too much on his plate to consider the report which he had written on the SAS, that it was something which the army was not prepared to consider at the time and would he please leave him alone. Beckwith began writing to Congressmen and Pentagon officials over the heads of his superiors and things got so bad that during one of his many visits to Bragg, SAS Commander John Woodhouse had to take Beckwith aside and tell him, 'Cool it, Charlie. I'm sympathetic to you but you've got to live here and we don't. What I'm saying, old chap, is that you want to take your time about this. You are going to catch more flies with honey than by batting them around.' Then Vietnam exploded.

Despite thinking that Gritz should in theory not have been allowed into the special forces, Beckwith wanted him to go back to the Cambodian border to recce the infiltration routes of two North Vietnamese Army (NVA) regiments which were besieging the Green Beret camp of Plei Me. Beckwith was going to be leading a relief force into the camp while Gritz guided in an Air Cavalry assault to cut off the North Vietnamese. Gritz had barely been briefed on his mission before he was on a helicopter with a team composed of two other Green Berets and three Chinese 'Nung' mercenaries.

As soon as they had landed, Gritz's team started spotting NVA logistical facilities, including a field hospital with 600 wounded. He relayed the information to headquarters, recommending that he act as pathfinder for a surprise attack by the Air Cavalry. Twenty-four hours later, Beckwith was on the radio, saying, 'Prepare for immediate extraction from your location.' Gritz protested that the Air Cavalrymen who were rookies on their first mission in Vietnam would suffer unnecessary casualties if they walked on NVA positions without adequate guidance. But Beckwith was adamant: 'You do as I say – prepare for immediate extraction.'

On the way back to base, Gritz's helicopter was ordered to divert to Plei Me, into which Beckwith had managed to manoeuvre his relief force around well-entrenched North Vietnamese positions. They touched down amid exploding mortar shells to find Charlie standing in the middle of all the action holding a grenade launcher and chomping on a cigar. The NVA resistance around the camp was so fierce that a counter-attack which Beckwith had led with two companies had been repulsed. North Vietnamese soldiers had been found chained to their machine-guns. Four American fighter aircraft and two helicopters had been shot down while providing air cover. Beckwith wanted Gritz to take out as many dead bodies as he could fit into his helicopter. Their stench was terrible. The cadavers were laid out in front of him wrapped in parachutes, swollen from decomposing in the heat. 'I fucking promised them a body bag,' said Beckwith, spitting out the butt of his cigar and disappearing back into the fray.

When Gritz got back to the tactical operations centre of the 1st Cavalry Division to deliver his full report, he asked the general officer why his Delta team couldn't act as pathfinders. 'Do you think I'm going to place the lives of my men in the hands of a single Special Forces captain?' came the reply. 'Our standard operating procedure requires any landing zone to receive one hour of aerial bombardment and thirty minutes of gunship suppression before the first troops are inserted – do you understand that? . . . you guys would have been in the way.'

Gritz was to find out later that the gunship suppression had destroyed the NVA hospital which he had located, leading to accusations that he had directed fire onto an enemy hospital. The Air Cavalry ended up fighting one of the bloodiest and longest single engagements of the Vietnam War, suffering hundreds of casualties.

'The aerial bombardment had only alerted the North Vietnamese to the impending attack,' observes Gritz, 'giving them plenty of time to take up hidden positions in the jungle from where to ambush the inexperienced Air Cavalrymen.'

The concept of Air Cavalry or helicopter-mounted infantry had been another creation of the Kennedy administration as it prepared the Army to fight highly mobile counter-insurgency wars. After the bloody nose it received in its first engagement, the 1st Air Cavalry developed its own Ranger companies to conduct small unit infiltrations and reconnaissance missions. The 75th Ranger Regiment, which considered itself a descendant of World War II's 5307th, with the letter 'M' for Marauders on its shoulder patch, was trained and activated in Vietnam.

In one operation a five-man Ranger team was inserted to conduct surveillance of an area of jungle suspected of concealing a Viet Cong base. The UH1H helicopter dropping off the patrol circled around the area, coming down again for two other false insertions to confuse possible enemy observation, while Sergeant McConnell started moving for several miles through thick bamboo vegetation with trees thirty to fifty feet high and three to five yards apart. By nightfall they had located a recently used bunker complex, and set up an ambush position to monitor the bunkers and trails.

In the darkness the patrol leader observed three flashlights and the silhouettes of twelve individuals approaching their position. When the Viet Cong point man, at the front of the group, came to within five feet of him, McConnell fired a burst of his M-16 rifle, killing him instantly. Almost simultaneously, specialists fourth class Leesburg and Burch shot another two. As the rest of the enemy fell back towards a stream, the Rangers fired at the end of their column, then worked slowly towards the front. Directed in such a way the heavy fusillade of automatic rifles and grenades fired from M-79 launchers gave the enemy the impression that they were encountering a much larger force. Moving to try to outflank the Rangers, the VC ran into a line of claymore mines which McConnell had placed in anticipation of such a manoeuvre. After suffering several more casualties, the enemy withdrew, dragging away their wounded.

Drawn in by the ambush, a large VC force which was now taking positions on the opposite side of a stream began pouring automatic and grenade fire on the five Rangers. McConnell called in reinforcements of artillery and suppressive fire from helicopter gunships. The dark jungle became illuminated with flares casting a greenish glow as artillery began falling on the VC. Standing up to get better radio reception, McConnell was hit in the arm. At that moment another Ranger caught sight of a VC crawling towards him and shot him in the face. The medic rushing up to attend McConnell was hit in his side. A Ranger's rifle jammed as he tried to shoot two more VC creeping up on him. Just as they were going to fire, another Ranger cut them both down in a long automatic burst from his M-16. A grenade then exploded, with several fragments wounding one of the Rangers.

It was 6 a.m. when the helicopter gunship came hovering overhead, firing rockets on the VC. The wounded Rangers identified their position with a strobe light. A Vietnamese interpreter speaking on a megaphone from a helicopter called on the VC to surrender. Cries of 'Chu Hoi' immediately went up from surrendering Vietnamese as McConnell tried to direct the gunship fire away from them. At 7 a.m. an Air Cavalry platoon linked up with the Rangers, then interrogated the Viet Cong prisoners and searched the bunker complex, uncovering documents which provided information on the Viet Cong network in the entire Phuoc Vinh Valley of South Vietnam.

Sergeant Oliver Rael, the son of a Pueblo Indian chief from a village of only sixty Indians in the mountains of New Mexico, volunteered to join the 101st Screaming Eagles Airborne Rangers in order to challenge himself and prove his patriotism. 'I felt that if I was going to be a warrior, I should go all the way.' He was sent on missions in which five-man teams were dropped into VC territory 'to kill, destroy and terrify and then radio to be lifted out'. Carrying two days of rations and 150lb of equipment, Rael would go armed with a 50-calibre machine-gun which he was trained to fire 'across a torso to slice enemy bodies in half'. They would dissect the anuses of dead enemy soldiers 'because our commanding officer believed that they carried microfilm and documents hidden there'. On many occasions Rael's patrols would cut off the testicles from dead bodies, then stuff the severed organs in the corpses' mouths and hang them from trees to create fear and demoralize other VC.

After a raid on one VC camp, Rael spent an entire night hiding under the overhang on the edge of a jungle stream and watched the barrels of the enemy rifles glinting in the light above him as the VC cradling their weapons sideways passed by directly overhead. He began to loose enthusiasm for the American war effort after finding the naked, bullet-riddled body of a Vietnamese prostitute lying atop a dead fellow Ranger on a bed in a brothel. 'If they were willing to kill their own women to get at us, then I guess they felt more strongly about what they were fighting for than we did,' says the five foot five, enormously broad-shouldered American Indian, who from an early age had learned to live off the fat of the land, hiking barefoot in the south-western Rockies and eating locusts.

Rael became completely demoralized during R&R (Rest and Recuperation) in Hong Kong when he watched his best Ranger buddy shoot up heroin. After calling his commanding officer to say that he wanted out, Rael was told that anyone else would have been court-martialled, but since he had performed with such courage in the past he could still get a quick honourable discharge if he returned to Vietnam to train new Ranger recruits in patrolling tactics. During one session in which a new group was divided into three patrols learning how to set up booby-traps, Rael listened on his walkie-talkie as Patrol A walked right into the trip-wire set by Patrol B and screamed in agony as the claymores exploded. Before Rael was able to stop the exercise, Patrol C came running to their rescue and stumbled over the same trip-wires shredding their bodies with the shrapnel exploding at waist level in a 180-degree radius.

'Bo,' whispered Colonel Francis 'Blackjack' Kelly, the commander of US Army Special Forces in Vietnam, 'when are you going after that box?' The huge man held Gritz close to him with both arms like a kindly father, talking quietly into his ear. 'What box?' replied Gritz.

'What do you mean, what box?!' said Kelly, pushing him away. 'The black box of course! When are you leaving to go find it?'

Gritz had left Project Delta after getting into a fight with Beckwith and was now head of the Mobile Guerrilla Task Force 957, being briefed on his next mission. 'Bo could never get it right,' muses Beckwith. 'He never understood what reconnaissance was all about. Every time that he went out on patrol, he always had to shoot somebody.' Even in a war in which a mission's success was generally measured by body counts, Gritz set an impressive record: 400 KIAs (killed in action) by the end of his two years.

But Gritz's explanation for his falling out with Beckwith has little to do with shooting. Having become Delta's executive officer, Gritz was one day listening to a young soldier who did not want to go on another mission because he was quite frankly scared. His nerves were shattered and he explained that he had had a premonition about dying. Gritz understood this and ordered him to be transferred to another unit. But when Beckwith got wind of it, he exploded, 'Court-martial the bastard for cowardice!' As Charlie readily admits, 'I do get overemotional at times.'

There was no question about Gritz's willingness to carry out whatever it was Blackjack Kelly wanted him to do even if it meant the impossible. He just hadn't been briefed on his mission. Turning on his staff officer, the angry, intimidating, hard-faced, six foot three inch, sixteen-stone New York City ex-cop bellowed, 'Didn't you give Gritz the black box mission, like I told you?'

'Sssss . . . sir,' the frightened major was finally able to get the words out. 'We knew how important this mission was but we didn't think it wise to divert Bo to go search for a needle in the haystack.'

'Don't talk to me about needles in the haystack! You get your ass in the commo bunker and get a briefing team up here now!'

A U-2 spy plane on a secret mission over North Vietnam had malfunctioned at an altitude of 70,000 feet and crashed into the jungle. Its pilot had ejected and been recovered but the plane's sensitive 'black box', containing a top-secret electronic jamming device designed to deceive enemy radar about the position of U-2s, could have survived the wreckage. If it fell into North Vietnamese hands and was passed on to the Soviets, the entire US strategic air reconnaissance programme could be seriously compromised. The White House and the Defense Department had ordered the black box to be recovered at all costs. Gritz's mission was to locate it in a square of jungle with an area of some 625 square miles on the border with North Vietnam which was under total VC control, and to bring it back.

Inserted by helicopter with a twelve-man A team and 100 of his Cambodians, Gritz hacked his way for three days through dense forest where supplies had to be discreetly dropped by parachute in napalm canisters by fighter planes pretending to be on bombing runs. For any unusual helicopter activity would alert the Viet Cong. When they finally found the wreckage of the U-2, 'the Pratt & Whitney engine was partially exposed so that some of the compressor blades were visible. The Cambodians and the Green Berets climbed over the bird as if they were kids in a McDonald's playground.' But the black box was missing. 'The Viet Cong had been there. BF Goodrich sandal tracks were everywhere and the tail section of the plane containing the ECM 13A electronic jamming system was missing.'

The only thing to do, Gritz decided, was to take some VC prisoners and extract information about where the black box had been taken. They dug in along a well-used jungle track and set up an ambush. But on the first try, the Cambodians were overenthusiastic, killing all of the ten VC that they surprised. Gritz sent most of them back to their bivouac area and carried out the next ambush with just his A team. When a patrol of eight VC walked into the trap, gunfire devastated the six at the rear while Gritz and one of his sergeants jumped out of the jungle to take the two in front with their bare fists. A six-degree black belt in karate, Gritz unintentionally killed one of them with a single blow. But the Green Beret sergeant managed to take the other alive. Marching the prisoner back to camp, Gritz convinced the badly beaten sixteen-year-old, who had a dislocated arm, that he would die of his injuries unless they got him to a doctor, but where was the tail section of the plane?

'It in our camp, five miles from here,' came the prisoner's ready reply. 'I show you.' They approached the VC base while it was still light enough to be able to search for the black box once they had fought their way inside. The VC guide recommended that they approach through the latrine area at the back which was generally unguarded because of the stench. Pouring in with all guns blazing, Gritz's Guerrilla Task Force caught the VC entirely by surprise, pushing them into a defensive pocket at the far end of the camp while they located the U-2's tail section and detached the black box.

Gritz radioed in a helicopter to extract the black box and their prisoner as soon as he had disengaged from the raid. But after deciding to use up his remaining supplies and ammunition, he led his force through several more days of mayhem in the 'denied enemy area' before a full evacuation. He looks back on the black box operation as 'a side thing when I was trying to shake down an entire guerrilla army. But if it had been a high-level CIA operation without the simple integrity of army field officers, I could have requested $10 million in gold bars to barter with and had it dropped off with no questions asked.' As it was, Gritz got $150 for his troubles and was told to 'go have a party'. As it was top secret, no formal mention was ever made of the black box mission and no official commendation ever received.

Obeying their patrol commander's hand signals, all five men fell into concealed positions, forming a small, thumb-shaped perimeter in the thick jungle. Sitting up, they each slipped off their heavy packs weighed down with 200-300 rounds of ammunition and gently lowered them to the ground, then lay alert for several minutes with their rifles and machine-guns cocked, straining their five senses to pick up any sign of enemy presence. Amid the foliage which covered them, sight was the sense which they could least rely on. Yet all that could be heard beyond the zone of silence surrounding them were the usual animal noises: birds, a baboon in the far distance.

The men of the Australian Special Air Service Regiment were acclimatized to the jungle, having fought alongside the British in Borneo before coming to Vietnam. There they could cover as much as 10,000 metres a day. But here they were up against a much bigger and harder enemy. This was their second listening halt in an hour. It was impossible to see for more than a few metres in front. After a few minutes Sergeant Joe Van-Doroffelaar motioned with his arm. The man nearest him got on his feet and disappeared into the jungle. Hessian cloth strapped around his boots muffled the sound of his footsteps and blurred the traces of any marks as he moved slowly for about thirty metres, well beyond the viewing zone of the rest of his patrol. There was just more foliage and bamboo; and no distinct sound or any 'highway signs'. He returned to his position. With nothing to report, there was no point in even wasting a whisper to Van-Doroffelaar, who now motioned to the next man.

This time, when the corporal appeared back out of the bushes he pointed in the direction he had come from. Immediately Van-Doroffelaar signalled with a swoop of his arm for the rest of the patrol to follow him and soon they were all seeing footprints and sniffing the distinct odour of fresh human urine. They moved very carefully now, not even ruffling a leaf. Then they heard a noise directly in front of them, the clink of something metallic. Rifle butts went up to the shoulders, ready to fire before the second noise of a turning bicycle chain. Van-Doroffelaar moved forward about a metre and peering through a clump of bamboo, caught sight of two North Vietnamese soldiers in their green khakis, pushing their bicycles onto a well-travelled track. It was as close to a road as one could get when straddling Vietnam's border with Cambodia, and totally covered from the air by triple canopy jungle. This had to be the hidden route of the Ho Chi Minh trail which they were looking for.

What had originally started in 1957 as the Special Air Service company of the Royal Australian Regiment, had been reorganized into an independent Special Air Service Regiment in 1964 with the encouragement of the British, who needed help with their various counter-insurgency campaigns. Following a visit to the Australian SAS headquarters in Perth by the peripatetic John Woodhouse, the new special forces unit was deployed to Borneo. The Australian high command had at first resisted supplying the specialist hardware required for SAS operations: canoes, plastic explosives, commando knives, handcuffs and abseiling equipment, as well as unusual arms and munitions of foreign manufacture. The Australian SAS became notorious as a unit where the rest of the army sent its troublemakers. 'Most in the SAS at that time were types with chequered backgrounds, coming from broken homes and broken marriages,' recalls Bob Mawkes, remembering the constant fist-fighting when he joined as a trooper. 'Many had previously served in the French Foreign Legion. There were many Eastern Europeans.'

But despite its reputation, or perhaps because of it, the two squadrons composed of eighty-six enlisted men and three officers soon began setting records. Sergeant Arch Foxley led the longest jungle patrol in history, surviving for over three months deep in the forests of Borneo. An Australian four-man team conducted one of the first 'claret operations', sinking river barges carrying supplies to an Indonesian military base. The Australian SAS fired some of the last shots of the war in Borneo on January 1966, only a week before the leftist government of President Suharto was overthrown in a military coup led by General Sukharno, who negotiated a peace settlement acceptable to Britain and the USA.

During their last action in Borneo three Australian SAS men who had burned themselves with one of their phosphorus grenades while trying to take an Indonesian machine-gun position, managed to charge through the encircling enemy and evade capture for an entire day until reaching a British Army base on the other side of the border. But the Australian SAS suffered its most serious casualties of the campaign in an encounter with a wild elephant which attacked a four-man patrol. Proving as invulnerable to their 7.62mm bullets as a tank, the rogue beast trampled on one of the soldiers and gored him with his tusk, then picked up another with his trunk, hurling him against a tree. Both men died of their injuries.

Augmented with a third squadron, the Australian SAS was committed to Vietnam as part of Australia's contribution to the war effort through its partnership with the USA in the South-East Asia Treaty Organization (SEATO). SAS units arrived ahead of the 6th Royal Australian Regiment, and were quartered in camps south-east of Saigon which at first consisted of nothing more than flooded foxholes. During the battle of the Long Tan rubber plantation, they rolled out into the jungle from the back of armoured vehicles to hunt down VC units and gather sufficient intelligence to allow the 6 RAR to root out an entire VC Division from the area. The Australian SAS soon became known to the enemy as 'the jungle ghosts'.

SAS men were attached to American special forces units throughout Vietnam and the pentagon appointed a study group to investigate and try to assimilate SAS patrolling methods. 'We could teach the Americans some of the tricks of the trade,' says Joe Van-Doroffelaar. 'Their map reading was not quite up to our standard. They tended to rely too much on photographic maps. But their biggest problem was noise. They tended to break the code of silence too easily, breaking into chatter and smoking while on patrol. They also smelled of body deodorant. It was the reason why a lot of American reconnaissance patrols were compromised. A good jungle operator has to know all the natural noises and smells of the jungle and do nothing to alter their pattern. If necessary, a patrol should go without talking for days.' During their seven years in Vietnam, the Australian SAS suffered only one casualty to enemy fire, Sergeant George Baines, whose patrol got ambushed approaching a major enemy camp. 'We initiated 90 per cent of all our contacts with the VC.'

As soon as they were on slightly elevated ground, the five Australians formed an ambush line about ten metres from the road. Unpacking the two claymores which each carried in his rucksack, Van-Doroffelaar and another man crawled forward to plant six of them by the edge of the jungle track, standing up the curved boxes filled with layers of plastic explosive on their pincer-like miniature bipods. After covering them in leaves, they placed three other claymores to screen their flank. One was kept in reserve. Belts of 7.62mm ammunition wrapped in plastic sheets to protect them from moisture were passed down to the machine-gunner with the heavy 20lb M-60 dismounted from an armed helicopter. The other men laid out extra ammunition clips for their British L1A1 self-loading rifles within arm's reach while Van-Doroffelaar made sure that his Soviet-made Kalashnikov AK-47 was on full automatic to avoid releasing the clearly audible click of its safety-catch at the wrong moment. The compromising noise which he found to be the only defect in the VC's standard weapon had led to many of the twenty confirmed kills he had scored so far in Vietnam.

General Westmoreland, the American commander of land forces in Vietnam, had strictly forbidden the use of captured AK-47s. But being among the most experienced jungle patrol leaders, even at his young age of twenty-one, Van-Doroffelaar knew the reason for such a ruling. Copying SAS tactics in Malaya, the CIA was 'jarring' thousands of the rifles with a sprayed-on explosive which blew up the weapon in a user's face and planting them in VC arms caches. Van-Doroffelaar also knew that there was nothing wrong with his AK, which was lighter, more rugged and easier to handle than the British SLR or American M-16 and better suited to the close quarter combat of the jungle. Special forces had the privilege of choosing their individual weapons and Van-Doroffelaar liked to tell his men that 'happiness for an SAS soldier is a warm AK-47 and two KIA before breakfast'.

It was around breakfast time the next morning that the Australians lying in ambush finally heard some heavy movement coming their way. More than one had the sticky sensation of excrement running down his leg as they were shaken awake. The scout who crawled over to report that a large North Vietnamese column was heading down the track took up a position behind the rest to provide flank and rear security. They all draped their netted scarves across their faces and lay perfectly still behind clumps of bamboo, adrenalin rushing through their veins as they anticipated the deadly moment.

There were pounding footsteps, many of them. Then they could hear chatter. Finally NVA soldiers started coming into view. It was a whole column, some thirty or so marching four abreast, carrying baskets with food and mixed supplies. Ammunition cans were suspended on poles shouldered between them; weapons were slung on their backs. A flag indicating that the group was a headquarters element flew in front. Van-Doroffelaar spotted the leaf-decorated pith helmets of the officers in his gun sights as his left thumb pushed the button of the detonating cord connected to the claymores and the carnage began.

Showered by shrapnel from the exploding mines, the entire flank of the column seemed to peel off as a furious volley of automatic fire ripped deeper in, turning the enemy troops into a reeling mass. Ammunition cans exploded, turning the centre lines into a pile of corpses. Grenades were tossed into pockets of survivors as the M-60 machine-gunner hosed his lethal spray down on a group trying to break off towards the rear. As the belt-fed M-60 rattled incessantly, denying the enemy a chance to recover their balance, the others reloaded their SLRs with a third or fourth clip for the overkill. Van-Doroffelaar shot the last burst from his AK into an officer who still seemed to be moving and then got on his feet to rush into the mass of cadavers and retrieve enemy documents. The smell of cordite filled the thick, humid air.

Van-Doroffelaar was just stepping back into the jungle cover with several leather pouches slung over his shoulder when bursts of automatic rifle fire suddenly rang out from within the jungle. The spray of bullets began ripping into the foliage around the SAS patrol. North Vietnamese troops who had undoubtedly been following the ambushed column were trying a flanking manoeuvre. The trooper providing flank security detonated the second screen of claymores as the enemy's movement became visible in front of him. The radio man tuned into the frequency of a nearby helicopter to request extraction. Van-Doroffelaar motioned his men to begin moving towards the designated extraction point. The M-60 machine-gunner covered them, laying down a thick spray from his remaining ammunition belt before running after his mates, who in turn halted, firing their SLRs to cover his retreat.

The patrol kept moving – fast. As the sounds of the approaching North Vietnamese seemed to be closing in on them, the Australians threw phosphorus grenades, which let out their blinding streaks of white light and flames through the dark jungle, screening the SAS men behind thirty seconds of heavy white smoke. They passed a small clearing and Van-Doroffelaar ordered a pause while he checked a compass to get his bearings. There were the sounds of breaking leaves again as the North Vietnamese continued in pursuit. He ordered the last claymore to be rigged up in a bamboo thicket with a booby-trap wire strung across their track. They moved out as more shots rang out and half a minute later the exploding mine could be heard along with enemy shrieks. 'That will get them off the habit,' Van-Doroffelaar couldn't help muttering.

A helicopter was hovering sixty feet overhead as the five men approached the jungle clearing for the designated landing zone. But as they moved through the thinning trees, they met a barrage of automatic fire. The North Vietnamese had correctly guessed their extraction point. Thick nylon ropes weighed down by sandbags were flung from the helicopter. The enemy soldiers started coming into view as the SAS patrol ran into the clearing to hitch themselves onto the cables. Bullets came within inches of them, grazing one arm. Undeterred by the helicopter's machine-gun fire, North Vietnamese soldiers charged into the landing zone and Van-Doroffelaar emptied his AK into them, firing it with one hand as he was winched into the air. 'It was one of my eight hot extractions during my two tours in Vietnam,' which this son of a Dutch submarine commander describes as 'the best years of my military career'.

Captain Barry Petersen was having a difficult time holding down his rice wine. Like other special forces soldiers operating with the Montagnards, the aboriginal hill tribesmen of western Vietnam, the Australian officer was being initiated into his village through a ritual which required him to drink a slaughtered buffalo's blood washed down with what seemed like gallons of rice wine. It was one of the pleasures of gaining 'hearts and minds'.

Wearing a tribal *rhade* jacket edged with bamboo, a loincloth and beads, Petersen was trying his best to be polite to his hosts. They passed him another earthenware jug with a straw, insisting that he drink more of the glutinous stuff which had been fermenting for months. He had been taught at special forces school that it was a matter of etiquette never to refuse anything which a foreign 'asset' may offer as a gift or in the way of nourishment. But Petersen was about to be sick.

Petersen would spend several years operating in the critical highlands spanning the border between Vietnam and 'neutral' Cambodia, organizing a defensive network among the local tribespeople to seal off North Vietnamese infiltration routes. Since their early years in Vietnam, American special forces A teams had been building 'fortified villages' among the Montagnards, a distinct ethnic minority who had always resisted domination by central Vietnamese authorities. But Petersen, one of a handful of Australian special operatives brought into the sensitive mission, considered the Americans' strategy to be cumbersome and ineffective.

'It took fully a third of locally raised Civil Irregular Defense Groups to keep the forts with their machine-gun posts, observation towers and mortar pits permanently manned, generating a siege mentality and cutting down the manpower available for patrolling.' When a fortified village came under attack, the tactic of reinforcing it with CIDG units from other villages only played into the hands of the Viet Cong. They would set up an ambush of the relief force rumbling through predictable routes in American-supplied lorries. The besieged 'fortified village' became something of a cliché of the Vietnam War. More than one congressional medal of honour was posthumously awarded to Green Berets fighting desperate and gallant defensive actions.

In one engagement Captain Roger Donlon, was commanding officer of US Army Special Forces Detachment A-726 at camp Nam Dong when a reinforced Viet Cong battalion suddenly launched a full-scale pre-dawn assault. With the initial onslaught, he swiftly marshalled his forces and ordered the removal of needed ammunition from a burning building. He then dashed through a hail of small-arms fire and exploding hand-grenades to abort a breach of the main gate. Although exposed to intense grenade attack, and despite sustaining a severe stomach wound, he succeeded in reaching a 60mm mortar position. On discovering that most of the men in the gunpit were also wounded, he directed their withdrawal to a location thirty metres away. Donlon was carrying the mortar as well as dragging his wounded team sergeant when an enemy shell exploded, inflicting another wound on his left shoulder. He then retrieved a 57mm recoilless rifle which was in danger of falling to the attacking VC and received a third wound in his left leg. After crawling 175 metres to an 81mm mortar position, he directed fire to protect the seriously threatened eastern sector of the camp. He kept moving from position to position, hurling hand-grenades at the enemy until a mortar shell exploded next to him.

The heroism displayed by Donlon and other Green Berets was epic. But Barry Petersen was more interested in beating the VC at their own game. He had carefully studied Mao Tse Tung's book on the People's War, which described guerrillas 'swimming like fish in the sea of population', and tried to apply these concepts in leading his CIDG unit his 'Tigermen'.

Petersen didn't bother setting up static defences. He kept all of his 350 men continually on the move, patrolling from village to village, gathering intelligence and tracking the enemy. Ammunition and supplies were hidden in small caches and food was provided by the villagers. He made sure that men from different villages were mixed together in the different patrols so that the units could develop their own identity and at the same time have as much information flowing to them from as many locations as possible.

When the VC entered a village, Petersen just let them. 'Finding the settlement exposed and ungarrisoned, the enemy would not concentrate a very large force. They would undoubtedly commit some atrocity like killing the village chieftain and cutting off his wife's breasts, creating more local converts to the anti-communist cause. The VC would also have to get out again and that's when we got them.' There would always be a patrol nearby, waiting in ambush.

If it became necessary to retake a village, Petersen would place it under observation. On determining that resistance was light, he would surround it with eighty-four men spread out in groups of three. Gradually they would close in and the encirclement would be barely noticeable before the Tigermen were practically on top of the Viet Cong. 'People ran for the trenches which they had been ordered to build. The more intense firing could be heard from the far side of the settlement. Some Viet Cong trying to escape our assault force had run into our team lying in ambush around the outer flanks. A shot came close from a clump of trees. I immediately fired a short burst back into the clump. Silence fell as I cautiously approached the trees. Lying crumpled among the trunks, I saw the VC I had badly wounded. He was obviously dying and no longer a danger. I picked up his weapon and continued on my way. From behind, a shot rang out. The commander of my training cadre, Y-Buy, an ex-VC himself, had killed the wounded man.'

'I finally threw up my rice wine during my Montagnard initiation,' confesses Petersen years later. We were having drinks at the Madrid Bar on Bangkok's Patpong Road, a meeting place for old Indochina hands nestled among the go-go joints and brothels of Thailand's main red-light district. He learned to love South-East Asia and, like a few others, stayed behind after the war. The Australian's guerrilla tactics were exactly what, back at the John F. Kennedy Special Warfare Center, General Yarborough had been trying to teach his men. '. . . you have to accept enemy penetration for a while and apply radiation therapy for a while. It's slow, it's unpleasant, it's painful and it's not going to cure you quickly.'

While Captain Petersen lived among the Montagnard mostly on his own for several years, the American twelve-man A teams were rotated out every twelve months, living in compounds separate from the villagers. When he occasionally dropped in for a cold beer, Petersen would see the *Playboy* centrefold which the Green Berets had pinned on the wall. Miss December's luscious body was criss-crossed into 364 spaces with the last box to be blacked out, on the day scheduled for their departure, being 'pussy'.

'You said you wanted volunteers for an especially tough mission, sir? Well, here we are, ready to go. What's up, mate?' said the six foot four, seventeen and a half stone Captain Stanley Kraznoff of the Australian SAS, reporting for duty to Bo Gritz. Having been promoted to major since the black box mission, Gritz was now organizing another Mobile Guerrilla Force (MGF) consisting of his Cambodians, Green Berets, Rangers and a five-man team of the Australian SAS headed by Kraznoff, the son of a Russian Cossack officer who had emigrated to Australia shortly after the Bolshevik revolution. When the Australian liaison officer in Saigon had asked Gritz what the chances were of the men ever coming back, his honest answer was 'fifty-fifty'.

'The Aussies added both humour and combat power to our task force,' says Gritz, who appointed the SAS's irreverent Sergeant Major Sonny Edwards as 'my top kick'. While his job was to exercise close control over the often unruly troops, Edwards would never salute an officer without saying 'Pug Mahone', Gaelic for 'kiss my arse'. Gritz had to explain to visiting VIPs that it was Cambodian for 'Freedom and Work'.

Soon they were off together rescuing a CIA operative who had been captured by the North Vietnamese, or taking enemy prisoners in Cambodia. During one operation Gritz found himself taking cover in the same ditch as Edwards as they prepared to pounce on a North Vietnamese patrol. 'We lay on our backs, foot to foot, pointing our silenced 9mm sub-machine-guns just above our heads. The enemy came into our sights and I shot the first in line while Edwards took out the ones in back, leaving the officers in the middle for capture. Since missions generally came back to back, the MGF for a long time managed without a permanent base. Gritz called them his 'fighting gypsies' because they moved around in their fleet of helicopters from one launch site to another.

They were all kicked off the air base of Bien Hoa when, after getting drunk during a strip show by leggy Australian girls, they gatecrashed a dinner party being held in honour of the South Vietnamese premier, Air Marshal Nguyen Cao Ky. During the ensuing brawl, in which military policemen trying to encircle the Australians got badly mauled, the steel-jawed Edwards broke a leg pressing his assault on a jeep speeding away with some of the terrified MPs. 'If the VC had only fought more battles in bars, with the quick-fisted Aussies we would have won easily,' observes Gritz. As it was, an angry general told Gritz the next day to 'Go back to the jungle where you belong'.

In November 1968 Vietnam's lunar new year found Gritz and Kraznoff fighting their way through the narrow streets of Saigon. Documents which they had uncovered during an MGF reconnaissance raid on a VC hideout some months earlier indicated a growing mobilization by the communists which had now culminated in the Tet offensive. Pitched battles had erupted in every town and city throughout South Vietnam, and in Saigon even the US Embassy was under siege. The MGF had been given orders to clear out communist positions in the capital's Chinese district of Chon Lon. But Gritz's group was pinned down by heavy fire from a North Vietnamese machine-gun position.

'Let me do it, major.' On hearing this, Gritz turned to look at Kim Jane. She could never quite manage to do up the top buttons of her tiger suit and as she leaned over, her firm cleavage came into view. It was a wondrous panorama which would grab a man's attention even as 3000 rounds a minute flew overhead.

Kim was a French-Vietnamese stripper whom Gritz had met while on R&R at the coastal resort of Vung Tao. She had used her influence with the local authorities to find a beach-side base for his MGF and was brought into the unit as a 'nurse'. Gritz now thought back to their first night together. His first reaction had been to reach under his pillow for the 9mm pistol and the stiletto knife which he always tucked away in the event of unexpected nocturnal visits. His sixth sense detected a moving presence inside the tent which he shared with two other men who were fast asleep. 'Suddenly I felt something move like a serpent between the cot and the mosquito net. I grabbed it to notice a woman's hand and wrist.' Then the beautiful face framed by long, red hair came into view and Kim's hushed whisper brushed his face. 'Please Major, I need make love.'

He thought of her perfumed baths, of how she dangled her legs from the open door of a helicopter when he took her on rides. Of how she once confessed to him about having shot a previous American lover who had deserted her, and about the marksmanship talents which she displayed on the MGF firing range. Now she was volunteering to eliminate the North Vietnamese machine-gun nest in a way that only she could.

Gritz looked over to Kraznoff, who had tried to rush the machine-gun some minutes earlier, only to be turned back by its well-directed heavy fire. The Australian, with his mop of blonde hair falling over his sweat band, didn't say anything. Gritz looked back at Kim, who just repeated, 'I go, Major.' Reluctantly, he nodded.

She stripped off her uniform, arching her naked breasts as she pushed back her hair to fit on a traditional Vietnamese *ao dai*. The tight dress enhanced her every curve. She parted the long slit running up the skirt and strapped an automatic pistol to her thigh. Without saying a word, she disappeared down a connecting alley and walked around some blocks, to come into view of the four North Vietnamese soldiers at the machine-gun nest. They looked in stunned silence at this apparition, licking their lips in anticipation. When she got within a few yards, Kim revealed her leg, pulled out her pistol and shot them all dead. She raced back to Gritz's position, bringing back the dead North Vietnamese lieutenant's Russian-made Tokarev pistol as a gift. 'She was an incredible gal,' he says.

'Hey, you asshole, I came here to kick some fucking VC ass and take some fucking VC names and that's exactly what I fucking did tonight. And if you don't fucking like it, then fuck you and all your fucking kind, you sorry shit-for-brains, cockbreath, pencil-dick, numb-nuts asshole!' Ensign Richard Marcinko of the SEALs was standing on his river patrol boat (PBR), replying to a superior officer reprimanding him from the dock for calling in an 'unauthorized' air strike. As it happened, Marcinko had just prevented a major North Vietnamese army crossing into the Mekong Delta.

At the time of the Tet offensive, Marcinko's PBR carrying his platoon of eleven SEALs, had pulled into the besieged riverside town of Chou Doc. Having just got back from running an unauthorized patrol into Cambodia, Marcinko plunged into house-to-house fighting with a Green Beret sergeant, clearing a street which had been overrun by the VC, trapping an American nurse called Maggie.

Outfitted in VC-style black pyjamas, headscarves, and with camouflage grease covering their faces and ammunition belts strapped around their chests, the SEALs kicked down the door to Maggie's house as their machine-gunner sprayed the second floor from where VC were firing. Three VC who ran in from the back door were instantly hit by bursts of M-16 as the SEALs and Sergeant Dix poured into the house and bounded up the stairs. They shot two more VC on their way up and then another couple outside Maggie's room on the third floor, before crashing through her door and pulling her out of the closet where she was hiding. After running back downstairs, stepping over the bodies of all the dead VC, they threw her into a waiting jeep and speeded away. One SEAL had to protect Maggie by lying on top of her all the way to the army compound outside the town. 'What a great pair of tits,' he said after the ride. 'That's my kind of woman.'

Following two combat tours in Vietnam, Marcinko would go on to form the Navy's elite counter-terrorist unit, SEAL Team Six, but not before the Viet Cong had put a price on his head: 'Award of 50,000 piasters to anyone who can kill First Lieutenant Demo Richard Marcinko, a grey-faced killer who has brought death and trouble to the Chau Doc province during the lunar new year.' They got his picture and the name 'Demo Dick' from an article in the American magazine *Male*, written about Marcinko when he was doing a publicity tour for the SEALs. Tired of watching the Army's Green Berets get all the publicity, some admirals had apparently decided that the Navy's top-secret special forces unit, whose men weren't even supposed to wear name patches, should get some attention too.

If the Green Berets were John Wayne, then the SEALs were suddenly Rambo. *Male* had Marcinko with 'Hollywood good looks' jumping out of a plane at 25,000 feet above the Mekong Delta carrying a 57-calibre recoilless rifle. 'It was an atrociously written piece of fiction,' according to Marcinko, and got him on the enemy's ten most wanted list. He suspects that anti-war activists in America fed the press reports back to the VC.

'The truth is that the Navy did not know what to do with the SEALs,' explains Ensign Ron Yaw, who also went on to command Team Six. Admirals and captains who were Annapolis Naval Academy graduates only knew about ships, aircraft carriers and submarines. 'They had no concept about conducting guerrilla warfare in jungle rivers and swamps for which we were trained.' It was difficult to get the idea across to the Navy's top brass as the highest-ranking SEAL at the time was no more than a lieutenant commander.

Initially SEAL Team One had been deployed from its base in San Diego, California, to an area south-east of Saigon to fight on board the jacuzzi-powered PBRs, armed with machine-guns, mortars and 40mm cannon, which patrolled the rivers of the Mekong Delta.

When SEAL Team Two, the 'East Coast SEALs', began arriving in Vietnam in 1967, they weren't at all content with the laid-back habits of their West Coast cousins. Types like Marcinko and Yaw weren't happy with the role of glorified floating garrison troops in a backwater of the war. 'We wanted to get into a situation where we could kill somebody,' says Yaw, a college swimming champion who had joined the SEALs because he was sick of school. 'Action was our prime motive in life. Every time we went on patrol we wanted contact. As a platoon commander, it was my responsibility to design a situation where that desire would be fulfilled.' The CIA provided the opportunity.

Sam Wilson was the new Chief of Staff at the US Embassy in Saigon, putting together an organization with the ambiguous title of Civil Operations Revolutionary Development Support (CORDS), co-ordinating American-backed military and civilian (i.e. CIA) functions down to the provincial level throughout Vietnam. This gave rise to the Provincial Reconnaissance Units (PRUs): groups of VC defectors formed in each province under South Vietnamese police control to provide timely intelligence and assistance for 'direct action' missions against the Viet Cong 'infrastructure'. The CIA handed out bagfuls of piastres to individual SEAL units to contact their nearest PRU and recruit a Viet Cong defector or Chu Hoi – Vietnamese for 'I surrender' – who would guide them to a lucrative target. This became known as the 'Phoenix Program' in which SEALs waged their own secret war in the rivers, canals and inlets of the Mekong Delta to 'capture or kill members of the Viet Cong shadow government to weaken its capability to support military operations'.

Each SEAL platoon would generally designate one or two men to handle the liaison with the PRUs, which required regular visits to local police stations and getting to know the police and intelligence officials who could get them to the right Chu Hoi. There was one dark-skinned Panamanian-born SEAL with a good capacity for languages who had acquired fluent Vietnamese and was especially good at this sort of work. Once he had found a defector who was prepared to identify a VC village or province chief he would report the information back to his platoon commander. Yaw would then scan further intelligence on patterns of enemy activity in the designated village or province: frequency of attacks on PBRs, VC propaganda and tax-collecting activities. He might even board a helicopter and make a visual air reconnaissance of the area. 'If I ascertained that the identified VC chieftain was capable of coordinating supplies and organizing for company-sized enemy attacks, I would launch an operation.'

His platoon gathered at 5 p.m. inside the air-conditioned Quonset hut serving as their command centre. Yaw briefed them on their objective, the nature of the mission and all the intelligence on the target which he had been able to gather. SEAL operations generally could count on very little support from regular naval intelligence officers who, according to Yaw, wouldn't even provide maps. After the planning session they would go for dinner. Hamburgers or T-bone steaks were washed down with cold Cokes and hot black coffee. 'We would be ready to haul ass by sundown.'

Gone were the emaciated, exhausted, disease-ridden and leech-bitten Marauders or SAS jungle patrols of Burma, Borneo and Malaya. These were healthy, well-fed, tanned, muscular Navy divers and swimmers loaded with 300–400 rounds of ammunition each and grenades bristling from their carrier vests, who boarded a patrol boat to go upriver and 'do SEAL Shit' – orders from above were rarely more specific than that. Some patrols might take along a trained German shepherd or Rottweiler. But for Yaw's mission that night, they needed to approach in total silence.

'We went to sleep on deck for the thirty-mile ride to our insertion point. When the boat slowed down for the last mile or so, we all woke up, made last-minute checks on our weapons and reapplied green and black streaks of camouflage grease to our faces.' Yaw dispensed with his floppy jungle hat and tied a black scarf around his head, pirate style. 'There are lots more things you can do with a scarf than with a hat' he explains. 'You can use it for bandage, to tie explosives to something or to strangle somebody.'

The only one who had not slept was the Chu Hoi. He was too nervous, knowing that if the VC caught him it would mean unspeakable torture and a terrible death with his testicles stuffed in his mouth.

The seven men slipped off their boat in pitch-blackness and into chest-high muddy water, slowly wading to the shore with their arsenal of weapons. Yaw carried a Swedish light machine-gun called a 'grease gun' because of the huge amount of bullets that slip out in a short burst. Four others carried the standard M-16 automatic rifle and the machine-gunner had his Stoner, a drum-fed weapon with lethal firepower – no well-dressed SEAL patrol left home without one. The point man, who took less ammunition for the demanding task of going in front, carried a pump-action shotgun.

'Getting onto the river bank was the scariest part because you never knew what awaited you there. It was perfectly possible that the VC may have been alerted to the approaching PBR and would be lying in ambush.' Even the inhumanly tough Marcinko remembers having to take deep breaths at these moments and think back to 'that big-titted schoolteacher from New Jersey I banged back in the Virgin Islands' to keep from falling apart. The point man went about twenty feet inland and then stopped while the rest formed a semicircle down to the water's edge. The rearmost man kept a close watch on the Chu Hoi.

They waited twenty minutes, listening, smelling, distinguishing normal sounds and odours from those that might not be; adapting and attuning their senses to the environment. Slowly they moved on, Yaw looking frequently through his starlight scope, an early night-vision device which magnifies the natural illumination from the stars and moon, ordering frequent halts. 'Maintaining the element of surprise was totally crucial. We could be hell on wheels but only had enough ammo for ten minutes of heavy engagement. It is only the combination of surprise and shock which would make a larger enemy force disperse when we hit them with maximum initial violence.'

It took Yaw's platoon half an hour to cover 100 yards. Once they got through the thick vegetation near the water, the landscape opened up on a rice paddy. A well-trained patrol moves and breaths like one living organism, expanding and contracting in relation to visibility. While the men stay close together, not more than a few feet apart in thick foliage to maintain contact without communicating verbally, they can spread out to 100 feet in an open space to avoid becoming a target. Only the machine-gunner walking at the back with the Stoner remained close to the Chu Hoi. Yaw had given strict orders, knowing that it was not unusual for the informants to become so overwhelmed with fear that they would run away. In the best of circumstances, the usually young Vietnamese boys had been pressured or bribed by the authorities. At worst, they could be double agents. When one US Army special forces A team found that their Chu Hoi was a VC agent, they shot him between the eyes and dumped his body in a river.

In three and a half hours the SEALs covered over a mile and a quarter. It was 3 a.m. by the time they started to observe the huts, which were supported on wooden poles and grouped inside a zone of thick vegetation. Yaw moved next to the Chu Hoi and motioned the patrol to close up around him. The Vietnamese pointed towards a hooch about fifty metres from where they huddled. The patrol now took up firing positions as the point man ventured into the darkness. Crouching over his shotgun, he moved through the village to recce the approach to the hut. Indicating that it was clear, he got into a covering position behind it.

Leaving two SEALs to provide rear cover with the Stoner and look after the Chu Hoi, Yaw made for the hooch with the other three men. The radio man and another SEAL covered Ron, while Yaw and his second in command, Chief Petty Officer Gallagher, stood either side of the door and prepared to enter. 'Special Forces want to go up to Father Death, meet Father Death regularly and punch him in the eye,' is how Yaw explains the anticipation of that moment. 'It's the thrill of facing a very dangerous situation and conquering it that gives you the ultimate high; whether you are jumping out at night from an airplane at 35,000 feet, locking out of the escape hatch of a submarine deep under water or going into a room full of bad guys.'

The moment in which Yaw pulled up the cloth covering the entrance to see eight men sleeping inside with piles of AK-47s stacked up high against the wall, he knew that it was going to have to be a kill. He had pulled VCs out of their beds before, tied their hands and just walked them out to the helicopter or PBR rendezvous. When the danger level was low he would even spare the life of whoever might be sleeping with the man. But the scene reflected in his red-lensed torch appeared to confirm what the Chu Hoi had told him: this was the command post for an entire reinforced VC battalion – only a few miles outside Saigon.

'I pulled out my pistol, walked over to the main bed, lifted the mosquito net and shot two guys in the head. Blood and brain matter splattered everywhere as Gallagher, who was right behind me, hosed down a bunch with his M-16. A VC leaping from another bed tried to push me as he made for his AK but I punched him in the face before he could grab his weapon and shot him. Then I recognized the figure of a naked girl dashing out of the doorway and as I moved to try and grab her, there was an explosion in the hooch. Time froze. I realized that a grenade had gone off. As the smoke cleared, I could see Gallagher bent over with fragment wounds all along his right side. I stepped towards him and the next thing I knew I was chest deep in water.'

Yaw had fallen through a crater which the grenade had made in the floor and landed in a lagoon of rainwater below the stilts. He had passed out momentarily and as he tried to crawl out of the water 'I could hear increasing gunfire all around me'. He took one step and fell down, another step and fell down, then saw one of his men coming towards him. 'I can't move,' Yaw screamed. 'Don't know why.'

As the other SEAL tried to pull him up, 'I felt a massive pain in my right arm. Grenade fragments were embedded in my armpit and I had lost use of my left foot because a fragment had gone through my ankle.

'I was given a shot of morphine to kill the pain and put Gallagher, who although wounded could still walk, in charge of the patrol. The radio man had called in for a helicopter evacuation and a second SEAL squad which had been inserted separately was laying on heavy fire at the other end of the village to cover our withdrawal. Three of us were wounded but we had to rush because it was only one hour before dawn. The pain in my left foot was excruciating and I had to limp as fast as I could, using a rifle as a crutch. The point man led us through a forested area and into an open rice paddy. We got to a hooch. There were five Vietnamese inside and we were in no mood to ask questions. Having to assume they were VC, we shot them.'

Yaw, another SEAL with a gunshot wound and the radio man, got inside the hut among the dead bodies. 'The five others, including the Chu Hoi, who was now issued with an M-16 and shooting as furiously as the rest of us, set up a security zone outside.' Pursuing VC were beginning to appear out of the jungle. The Stoner was down to its last ammunition drum when Navy Sea Wolf helicopter gunships began hovering over us pouring fire from their chain guns into the tree line.'

Owing to the severe loss of blood, Yaw's condition was worsening. He felt like he was starting to freeze up, becoming overwhelmed with fatigue and no longer able to think clearly or concentrate enough to fire a rifle. 'All my energies were just focused on trying to remain aware of what was going on.' The radio man had come into contact with Army UH1H Hueys which were coming to pick up the SEALs. But the pilots were refusing to land because of the heavy gunfire from the VC. 'Goddamn fucking shit!' Yaw yelled. Then, over the radio, he could hear the voice of a Sea Wolf pilot threatening to shoot down the Army choppers 'if you don't take our guys out . . . Either you risk the VC or you motherfuckers get a sure one from us, do you roger that?'

In few minutes one Huey was landing directly in front of the hooch, pouring tracers from its M-60 machine-gun at a wave of VC running into the rice paddy. 'It was daybreak as I was helped on board and we all piled in the chopper with gunfire all around. Overweighed with eight of us, the helicopter was having trouble getting back up. But then, shudder, shudder, flap, flap, flap, we took off.

'I was still conscious when they got me to a field hospital, but when they tore open my boot and I saw the condition of my foot sunken in the pool of blood which had formed at the sole, I just passed out.'

During his six-month tour in Vietnam, Yaw took twenty-five Viet Cong prisoners, of whom half were province chiefs. Like other special forces officers, he had come to realize that the war was unwinnable unless ongoing operations could be mounted to root out communist sanctuaries in Cambodia which served as the supply and reinforce-ment base for all the Viet Cong and North Vietnamese elements he was fighting in the Mekong Delta. 'We would kill a whole bunch and a bunch more came back. When we did succeed in depleting the VC leadership in a particular area, they were simply replaced by North Vietnamese army regulars.'

Mobs of anti-war protesters back home, or in what the special forces referred to as the 'unreal world', chanted 'Peace Now' and 'Hell No We Won't Go'. Newsmen like Dan Rather became stars by debunking the US war effort which was making them famous. Jane Fonda decided to play groupie to Ho Chi Minh instead of John Wayne and went to Hanoi. But Yaw hardly fits the stereotypical image of the traumatized Vietnam vet consumed by guilt and self-pity. For him Vietnam was 'great fun. It was the perfect environment for us'. He believes he speaks for most of his men when he says that 'if the war had gone on for another hundred years, the SEALs would have been happier than hell'.

4

THE TERMINATORS

Through the sights of his telescopic rifle, Sniper 1 of the Groupement d'Intervention de la Gendarmerie Nationale could see a clear shot into the forehead of the terrorist dozing off to sleep with a sub-machine-gun on the back seat of the hijacked school bus.

Sniper 2 was managing to keep his cross-hairs firmly fixed on a second gunman of the Front de Libération de la Côte de Somalie (FLCS), who walked up and down the aisle, glaring at the twenty-eight terrified French schoolchildren between ages of eight and twelve. Sniper 3 could aim directly into the back of the head of a third terrorist near the driver's seat, while Sniper 4 would have no problem getting several rounds into the one standing outside.

'*Un*' . . . '*Deux*' . . . '*Trois*' . . . '*Quatre*' . . . Each marksman gave his code in sequence, indicating that all four Gendarmes were prepared to open fire on the targets precisely eighty-three metres in front of them. They waited for the order to shoot from Lieutenant Christian Prouteau, the twenty-eight-year-old commander and founder of the French crack counter-terrorist unit. On February 1976 the GIGN was on one of its early missions, involving the highest stakes of any which the unit had executed so far for the French government.

For the safe return of the schoolchildren, their teacher and the bus driver, the FLCS was demanding the immediate withdrawal of French troops from the colony of Djibouti on the eastern Horn of Africa, where the hostage drama now unfolded. To meet any such demands with even a token gesture would be an unthinkable humiliation for the government of France while the massacre of children was an unspeakably dreadful prospect to contemplate. It was a recurring nightmare which had been plaguing the western world since the massacre at the 1972 Olympic Games in Munich.

Guerrilla warfare had by now ceased to be something which the

West was going to do to the Soviet bloc, as some of the early founders of special forces had planned. Since the American defeat in Vietnam, terrorism had become a growing threat posed by emerging communist powers and their Third World satellites; it was aimed not so much at the military infrastructure of Western countries but at the moral fibre of their societies. It was much easier to erode the authority of governments and press non-negotiable demands through the publicity of blowing up an airliner, machine-gunning a crowded restaurant, kidnapping a celebrity or hijacking a bus full of schoolchildren than by sabotaging a military base. It was the ultimate form of psychological warfare with which the West at first found itself totally unable to cope.

European security organizations had been caught entirely unprepared when a seven-man team from Black September, an offshoot of the Palestine Liberation Organization (PLO), scaled the fences of the Olympic village in Munich and seized eleven members of the Israeli track team, then demanded the release of Palestinian prisoners. Lacking a properly equipped force to storm the compound, the German authorities agreed to allow the terrorists to board a flight to a friendly Arab country. Germany's Federal police planned an ambush at the airport with hidden police snipers and armoured cars. But a lack of training in the kind of co-ordinated, split-second timing that allows for no margin of error, and was necessary to ensure the success of a counter-terrorist operation, led to disaster.

They were not frightened, confused and semi-panicked common criminals who led their hostages onto the airport tarmac, but committed, indeed fanatical guerrilla fighters, highly indoctrinated and trained in the latest techniques of commando operations at special camps in the Middle East and Eastern Europe by experienced military instructors. When the German police snipers opened fire on the terrorists, two marksmen who were unsure of their aim made the fatal mistake of hesitating, which allowed the surviving terrorists sufficient reaction time to detonate hand-grenades and kill the eleven young athletes. Two of the terrorists actually lived through the inferno. If the Germans can't do it, how can anyone? was the stunned reaction of security officials throughout Europe.

The ghost of Munich haunted the GIGN snipers lying ten metres apart, stretched along a berm of sand in the desert of Djibouti awaiting the signal from Prouteau, who struggled with his particular 'rules of engagement'. Fearful of a repetition of the 1972 massacre and unsure about the effectiveness of their new counter-terrorist team, the French authorities had given the GIGN commander strict orders not to open fire unless there was only one terrorist on board the bus. The rest had to be outside. But the FLCS had its own rules and during the entire morning in which the GIGN had been in position, the number of terrorists in the bus had actually increased, or rather fluctuated between three and eight.

From his concealed position beneath a palm tree, Prouteau had noticed another problem which the bureaucrats in Paris were choosing to ignore. While a passing camel, a donkey and a dog occasionally interfered with the sniper's line of fire, he had detected evidence of rather more sinister activity around the school bus. Through his binoculars Prouteau could see constant coming and going between the bus and a frontier post just a few metres away, heavily guarded by troops from the neighbouring country of Somalia. Ruled at the time by a radical Marxist government which was closely advised by the Soviet Union, Somalia was highly supportive of 'national liberation movements' in Africa and the Middle East and might be particularly receptive to the FLCS, which proposed a merger with Djibouti. The terrorists were in constant communication with the border guards, who seemed to be assisting them.

'It's impossible, Lieutenant. We have received assurances from the government of Somalia that they are not in any way assisting the terrorists,' was the reply Prouteau received from his superiors in the Interior Ministry when he reported his own observations. 'Shit!' he said to his radio man after his conversation with Paris. 'If we look closely at this problem, the administrative people are more dangerous than the terrorists.'

It was 3.30 p.m. The desert sun was beating down and the breeze from the nearby beach blew hot. Prouteau was on the radio again to Paris and after much insistence managed to get the rules of engagement changed to two instead of one terrorist inside the bus. Looking through his binoculars, he could see that there were still three. He let the field glasses hang on his neck as he wiped the sweat from his forehead and thought back to a year ago, to when he could have shot Carlos. Ilych Ramírez Sánchez, the international terrorist known also as 'The Jackal', leading two Arabs of the Popular Front for the Liberation of Palestine (PFLP), had bungled an attempt to hijack an Israeli El Al flight at Orly airport, Paris. Drawing their sub-machine-guns when challenged near the ticket counter, they cut down two policemen, took eight hostages, including a mother and her young boy, and locked themselves in the nearest toilet.

As negotiations dragged on over the terrorists' demands to be put on a flight to Egypt, Prouteau and a team of five Gendarmes took up positions around Orly's departure lounge. Three snipers were concealed along the second floor balcony while Prouteau and two of his men remained close to the lavatory. The government finally agreed to let Carlos and his PFLP team board an Egyptair flight with their hostages. Prouteau, dressed in civilian clothes, gripping his .357 magnum revolver inside his windcheater, closely followed the Venezuelan-born terrorist and two machine-gun-toting Arabs as they walked from the toilet to the airliner. 'The time to get them without endangering the hostages was when they were walking up the ladder to the aircraft in single file. I had Carlos in my sights but was not allowed to shoot. Until the moment in which the aircraft's door closed I kept hoping that the orders would change.'

Carlos would return to France some months later to conduct more terrorism and shoot two plain-clothes detectives who tried to arrest him, before making another dramatic escape with the assistance of the East German Embassy. He would go on to Vienna, where he held hostage a conference of ministers of oil-producing countries and got away with a large ransom to Libya. The GIGN sent a team to Switzerland to try to capture him again when he was spotted in the fashionable winter ski resort of Gstaad but once more Carlos eluded them. Prouteau had sworn to himself never again to miss an opportunity like the one he had lost at Orly.

Once again the sniper team indicated their readiness to fire. Prouteau picked up his binoculars. The FLCS gunmen's positions remained unchanged. Three were inside the bus and one was outside. He also had to worry about the two heavy machine-guns pointing from the Somali border post. Two GIGN snipers were positioned in reserve to deal with that possible threat. In addition, two platoons of the Foreign Legion were guarding his outer perimeter and were prepared to reinforce him with an armoured car. It was now or never. Prouteau had total confidence in his men and was not going to allow the failure at Orly to repeat itself. For him it was as much a precedent as Munich. Rules be damned. He took matters into his own hands. 'Zero,' came his signal to fire in three seconds. The marksmen wiped the sweat from their eyes. Their forefingers brushed the flat curve of the triggers . . . 'Fire!'

Instantly the brains of all four terrorists were blown out by a hail of 7.62mm bullets. Managing to keep their aim despite the heavy recoil of their FRF1 high-precision rifles, the Gendarmes kept shooting as long as the four heads were in view, working their weapons' hand-operated bolts with mechanical speed. The FLCS gunman standing outside the bus was perforated by four rifle rounds during the second or two in which he hit the ground. The one dozing off on the back seat took three bullets through the head. The one near the driver's seat and the one walking through the bus were each penetrated twice through the skull.

But then, what Prouteau had feared most – the machine-guns at the Somali border post opened up on his positions. Soon the bushes around the small fortress which had been concealing a company-sized reaction force also erupted with AK-47 automatic fire. The four snipers dug into the sand for cover as bullets stitched up dust all around them. Prouteau unholstered his .357 magnum as his two reserve snipers positioned further back took careful aim and fired into the machine-gun nests, hitting several Somali soldiers and silencing the MG-42s for the few seconds necessary to allow the other four snipers to reload and fire off another deadly volley.

The minute and a half it took for the Foreign Legion to appear seemed like an eternity, and Prouteau began to worry deliriously about the safety of the children he had come to rescue, who were now caught up in the intense crossfire. As soon as the armoured car began to move into view, he jumped up, revolver in hand, and led his eight men in a desperate rush for the bus. Except for one sniper who kept covering the rest with his rifle, the others discarded theirs and ran with just revolvers. They needed maximum speed and mobility to enter the bus, pull the children out and onto the Legion jeeps which were driving up behind them.

A few metres from the bus one of the GIGN men caught a glimpse of a man running in through the vehicle's back door facing the border post. The Gendarme jumped in through the front door to be met by a burst of sub-machine-gun fire from the terrorist, who had just made it inside to slaughter the hostages. The teacher and several of the children sitting at the front were hit by the spray of bullets, which forced the Gendarme to take cover behind the driver's seat. A Foreign Legion corporal then rushed in, managing to let off some shots from his assault rifle, which was not on full automatic for fear of hitting hostages, but he missed the terrorist. The Gendarme was able to regain his balance and take careful aim with his magnum. Holding it firmly with both hands, pulling back the hammer with his thumb and shooting it over the heads of the children, he drilled three bullets into the head of the terrorist, whose body crumpled onto the back steps.

Four more Gendarmes rushed into the bus, which was splattered with blood and stank of death and fear, with several injured hostages and the bodies of three dead terrorists lying around. Immediately they broke all the windows with their revolver butts and passed the bodies of twenty-eight traumatized children into the dozen jeeps pulling up outside. A full-scale battle was raging around them as the Foreign Legion, covered by their armoured car, closed in on the border post. One GIGN sergeant burst into tears as a little girl, wounded by the last terrorist's machine-gun spray, died in his arms. Four other wounded children and the teacher were put on the first jeep, which sped off to a field hospital. In minutes the evacuation was complete. Twenty-three children and the bus driver had been rescued uninjured.

The Somali troops pulled back from the border post, leaving behind twenty-eight of their dead, which allowed the GIGN to make a close inspection of the site. By counting the shell cases strewn all over the ground, they calculated that 2000 rounds had been fired without wounding a single Gendarme. About a dozen of the dead lay next to both the MG-42 machine-guns, where the GIGN snipers had concentrated their fire when the shooting started.

Prouteau suddenly noticed one corpse – the only one that was not black. He recognized the camouflage uniform on the muscular blond man lying dead next to a wall as distinctly Russian. Kneeling next to the man, Prouteau pulled out his papers, which identified him as a Soviet major, undoubtedly in the Spetsnaz, the special forces. 'It was clear to me that he had been directing the terrorist operation,' concludes Prouteau. But the French government decided to 'keep evidence of Soviet involvement in the Djibouti episode under tight wraps.

Driving an unmarked car, Major N. slowed to a crawl on his approach to the village of Culderry in Northern Ireland's County Armagh on a dark, cold, rainy evening in January 1977. He needed to keep the windscreen wipers going but switched off the headlights. Nothing should be allowed to betray an unexpected presence around the little community of stone houses and green hedges which was the home of Seamus Harvey, an IRA gunman responsible for several recent assassinations. Feeling the 9mm pistol tucked beneath his raincoat Major N. felt a certain reassurance.

The attention of Major N. and other SAS men already hidden around Culderry was focused on a car which had been linked to the recent killing of a British soldier. Driven back to the village, it was now parked on a side-street. They believed that Harvey would be using it that evening for another round of extreme violence. Sometimes the SAS can lie in ambush for days, even weeks; hiding in bushes, ditches, on rooftops, behind farm fences or inside covered foxholes, closely observing an IRA vehicle or an arms cache, waiting for the right moment to catch the terrorists red-handed, armed or, if possible, in the act. Under those circumstances, an 'ambush' has the full military implications of a wartime operation.

Major N. had been in Northern Ireland for almost a year, commanding the first SAS squadron to be deployed in the British Army's intensifying struggle with the IRA.

After successfully fighting off communist insurgents in Malaya and Borneo, Britain had experienced its own mini-Vietnam, resisting the encroachment of radical pro-Soviet regimes among its former desert colonies in southern Arabia, which lay perilously close to Europe's main oil deposits. As usual, the SAS was on the cutting edge of these remote conflicts, climbing over 6000-foot escarpments to root out guerrillas from inaccessible caves, leading mercenary armies of nomadic desert tribesmen or landing planeloads of goats to trade in the battle for local 'hearts and minds'.

It was during its undercover war in Aden which concluded with a British withdrawal in 1967, that the SAS first developed the gunslinging techniques of Close Quarter Battle (CQB), designed for fast and accurate pistol combat by small teams of gunmen in enclosed urban spaces. 'Keeni-Meeni' operations, named after a Swahili phrase describing the movement of a snake, were first conducted in the narrow warrens and bazaars of the ancient Arab port city, where the SAS countered bands of Yemeni-trained assassins. The special twenty-man team recruited for the job were trained to draw a Browning 9mm automatic pistol from the folds of an Arab robe, or *futah*, and fire six quick rounds into a playing card at a range of fifteen feet. The commander of the SAS, Lieutenant Colonel Peter de la Billière, who had cut his teeth as a troop commander in Malaya in the hunt for the terrorist leader Ah Hoi, went on to standardize CQB training for the entire regiment. When European security services rushed to form specialized counter-terrorist units following the 1972 massacre in Munich, the SAS was already a step ahead.

The SAS no longer had to venture out into the far-distant corners of the crumbling British Empire to fight guerrillas. 'The Troubles' which had exploded much closer to home in Northern Ireland, bringing sectarian strife between Catholic and Protestant communities, had led to a renewed terrorist campaign by the IRA in the mid-seventies. Invigorated under a young Marxist-Leninist leadership, the IRA was plugging into the growing international terrorist network. Along with other violent extreme-left groups in Europe, such as Germany's Red Army Faction, Italy's Red Brigades and Spain's Basque ETA, the IRA received training and arms from bases in radical Middle Eastern states and Eastern Europe, as well as from traditional Irish-American sympathizers in the USA. One of its biggest arms shipments, intercepted off the Irish coast in 1973 as it was offloaded from a chartered freighter, included five tons of Soviet-made automatic rifles, grenades, RPG anti-tank rockets and Semtex plastic explosives being sent from Libya to contacts in the IRA.

Sergeant Mick was part of an SAS team in Belfast when he witnessed the IRA bombing at a Protestant pub. 'It was the worst carnage I had ever seen.' The 80lb Semtex explosive device, planted next to a large window, 'had the effect of a huge claymore mine. People were decimated by flying glass, being cut to shreds by swathes from windows, glasses and mirrors'. Eight were killed and twenty-five wounded. 'Pieces of bloodied flesh, hair and scalp clung to the walls and ceiling. The sweet smell of blood mixed with the acrid odour of gelignite filled the room. It made me very angry but also conscious of my own vulnerability . . . even if you hate the IRA you had to admire them. They really believed in what they were doing.'

Major N. wanted the ambush of Seamus Harvey to go without a hitch. He had been planning the operation for weeks. The SAS was only starting to learn the ropes in Northern Ireland, working with a network of informants recruited through the Royal Ulster Constabulary (RUC). Since arriving in County Armagh in 1976, D Squadron had made its share of errors. While chasing IRA guerrillas responsible for the massacre of a busload of Protestant workmen, SAS soldiers had strayed into the Republic of Ireland, and were arrested by the Garda, the Irish police. When D Squadron finally did catch an IRA terrorist crossing back into Armagh by posting a surveillance around his girlfriend's home, they ended up shooting their unarmed prisoner when he apparently tried to escape while they waited for a helicopter pick-up.

Major N.'s radio crackled into life as one of his surveillance teams communicated the intelligence that Harvey was walking in the direction of the car carrying an object which looked like a rifle. The major drew his pistol. Seconds later, another message informed him that two unidentified men were following Harvey. Orders and warnings did not have to be given. The information was being relayed to Major N. as well as his men hiding close to terrorist's parked vehicle. They all knew what to do.

As Harvey prepared to open the car door, torches suddenly shone on him. He had a pump-action shotgun. The other IRA gunmen walking some distance behind managed to fire a few rounds at the SAS soldiers who appeared out of the shadows and emptied the thirteen-round clips of their 9mm pistols in a deadly hail of bullets. Harvey was cut down before he could do anything with his weapon. One of the other gunmen behind him was wounded but managed to escape along with the other. Major N. rushed over to see the blood marks, then turned to look at Harvey's bullet-riddled corpse oozing blood on the wet pavement beside the car. The killing of Seamus Harvey was one of the first SAS operations specifically aimed at eliminating identified IRA gunmen.

Similar to the SEAL 'Phoenix' mission in Vietnam, 'the overall brief is to destroy the IRA hierarchy', according to Sergeant Mick. 'More emphasis is put on capture. Most briefings are capture or kill but in some cases it's just kill. It's rare but in cases where there is little intelligence to be gained out of taking them alive, it's better to kill.'

Michael Rose is one of Europe's top military commanders and has recently headed the UN peacekeeping operation in the 'European Ulster' of former Yugoslavia. Sitting in his stately office at the Royal Army Staff College in Camberley, Surrey, Rose dismisses allegations about a policy to deliberately exterminate IRA terrorists as 'a load of rubbish'. He has just wolfed down a sandwich for his lunch in the downstairs reception room, whose walls are decorated with classic military oil paintings and crossed cavalry lances. He was hosting a lecture by Professor Clutterbuck, one of the foremost British academic authorities in 'low intensity conflict'. The professor was speaking to a group of British officers wearing their regimental kilts, about the prospects of a 'narco-terrorist' state being established in Peru by the Shining Path terrorist movement. 'However,' cautions Rose, 'if the IRA is made to believe that we are assassins and murderers, it's the best thing for us.'

It is estimated that over fifty IRA gunmen, a considerable portion of the terrorist organization, have been killed by the SAS since 1976. During SAS ambushes it is not unusual for the teams to shoot well over 100 rounds from close range into targets which have been carefully observed beforehand. In the kind of terrorist gang warfare which developed in Northern Ireland, British Army patrols frequently had to encounter individuals known to be connected with the IRA without being able to do anything unless there was a clear and immediate motive. Rose's deputy in the SAS at one time, Major C., tried for years to get Francis Hughes, a particularly fanatical IRA killer who was nonetheless careful not to go around armed. When they did finally catch him in the act of approaching a target with an automatic weapon, Hughes still managed to slip out of a deadly ambush laid down by the SAS, albeit with severe leg wounds. Eventually arrested by the RUC, Hughes died a martyr's death during the highly publicized Maze prison hunger strikes, scoring a significant IRA propaganda victory.

Sgt Mick would man observation positions (OPs) from inside the roofs of terrace houses in Republican areas of West Belfast. Usually, while carrying out a routine search of the home, a regular Army patrol would sneak an SAS team into the enclosed attic, which in many cases gave access to the roof space of adjoining houses. There they would live for days, defecating into plastic bags, eating cold rations and sleeping in turns without the families living below ever detecting their presence. While a suspected IRA arms cache which Mick had been observing was about to be raided by a combined British Army and RUC force, he noticed two men armed with AK-47s walking onto the roof of a nearby abandoned warehouse, preparing to snipe at the security forces. 'Without thinking about it, I just picked up my SLR, aimed through the opening we had made by removing a roof tile and shot them both through the head with two rounds. We were positioned much closer to the IRA snipers than to our own forces.'

Protracted OP work became the centrepiece of the SAS counter-terrorist strategy in Northern Ireland. Four men taking turns using peep-holes in the cramped space of an attic or enclosed between the brick walls of terrace houses, or sometimes buried in rubbish dumps on waste ground with decomposing dogs planted in front to keep anyone away, became the reconnaissance patrols of the urban jungle. 'The SAS has been through several phases,' says Rose. 'A jungle phase, a desert phase and with Northern Ireland we entered into our urban phase.'

SAS men also move around in unmarked cars, looking scruffy with long hair and jeans, passing for unemployed workmen pub-crawling in Catholic neighbourhoods, all the time gathering information. But this approach is more likely to arouse suspicion in the tightly knit Catholic communities. 'When British undercover men are suspected in Catholic pubs, they will be followed into the toilets and asked for identity papers,' says an IRA 'dicker', or informer. 'Our tactic is to give them as much rope as possible with which to hang themselves and when they are discovered they are killed.'

In 1974, when Captain Robert Nairac was overpowered by the IRA outside a pub while trying to pass himself off as an Irish field labourer, he tried to use his fists and all the martial-arts training he could muster to resist being bundled into a car and taken for interrogation and torture. They shot him and left his dead body in a back alley. Another SAS undercover officer managed to discover just in time that his cover was blown while working one of the Republican areas of Belfast. Having escaped with his life, he was immediately flown out of Northern Ireland, to a quiet posting in Oman.

The IRA dicker maintains that British undercover soldiers have been uncovered while trying to penetrate his organization on at least five other occasions which he knows about. The IRA insulates itself from planted informants or 'touts' through a system of tight compartmentalization by which members of different 'Active Service Units' (ASUs) are strictly prevented from knowing one another's identities. When 'James' joined in the early seventies after proving his worth by driving a burning bus crashing through the gates of a British barracks, he didn't know that his own sister was also a member of the IRA; nor did he find out for years. 'My own wife did not know about my IRA activity for several years until I was forced to tell her one day because her life was in danger,' he explains.

An SAS team on the trail of one of the IRA's most prolific bomb-makers decided that the only way to get a close look into his backyard workshop was from the top of an unused water-tower. They climbed in under cover of darkness and for the next two weeks silently scraped out several holes through five feet of thick cement to place cameras with zoom lenses at several angles, to take thousands of photographs of the demolitions expert rigging his deadly devices. At night they used infrared lenses to photograph him inserting sheets of plastic explosives into pornographic magazines, fusing these with light-sensitive back covers which would detonate the explosive when they were picked up. It was an effective booby-trap to plant among young British soldiers. The surveillance led to a well-planned capture by the RUC. They definitely wanted this one alive. His information would prove invaluable to Army bomb-disposal teams.

Elsewhere in Belfast, British special forces were rigging their own pornographic booby-trap but without explosives. Soldier F could hear a woman's moans grow more intense beneath his rooftop OP in the Republican Complaints Office. He half-lifted a floorboard and saw the owner of the pub below with the undressed wife of the local IRA chief, her legs spread wide as the publican eased himself into her. She was a real redhead and big, freckled breasts popped out of the open blouse as he fucked her on the landing at the top of the stairs.

Soldier F quickly reached for his video camera, zooming in on every detail of the steamy sex scene which went on for quite a while. 'The couple obviously got the most out of these stolen moments.' Some weeks later the head of the Republican Complaints Office received the film reels depicting his wife's X-rated debut and the Don Juan of a publican was never seen again. Tipped off, he managed to escape across the border before his boss's heavies caught up with him. For a few days the randy redhead was reported to be looking rather bruised and wearing sunglasses. Special forces operations also break hearts and minds.

The British Army headquarters which has grown at Bessbrook in South Armagh and from where most SAS operations in Northern Ireland are planned and co-ordinated, resembles a space city, thick with aerials and radars sticking out into the grey sky over acres of green countryside. It is the busiest heliport in Europe, with Chinooks and a variety of other helicopters taking off or landing every few minutes. Regular Army OPs are spaced out every 300 metres over the surrounding hills, stretching into the urban sprawl of Belfast, where soldiers occupy penthouse views on the tallest buildings.

SAS shoot-to-kill operations have continued until very recently, usually taking place around known IRA arms caches watched from the hidden close-up positions which are an SAS speciality. In a typical ambush, which took place at Christmas 1991, two IRA men picking up automatic rifles stored near a farm were surprised by undercover soldiers who appeared out of a ditch and killed both gunmen, one of them a well-known elected councillor for the IRA's political wing, Sinn Fein.

In February 1992 a six-man IRA team was surprised in the car park of a Catholic church after attacking an RUC police station. They were sitting in two cars dismantling their sub-machine-guns and an anti-tank rocket launcher when SAS men shouted a warning to freeze before pouring out 9mm bullets a second later. Four of the terrorists were killed and two were captured and later charged with killing several RUC officers and members of the Protestant Ulster Defence Regiment. The IRA arms storage site in the graveyard of the church had been under observation for some time. 'When you get the SAS involved,' says an RUC superintendent, 'you know that there are going to be very few prisoners.' A special RUC unit, E Squadron, is trained by the SAS in undercover surveillance and ambush methods.

The SAS Counter Revolutionary Warfare Wing (CRW), or 'Pagoda', grew in size and importance as the Regiment's involvement in Northern Ireland escalated and acts of international terrorism increased worldwide. Permanent liaisons and officer exchanges became established between the SAS and counter-terrorist units in other countries where the new kind of war blurred the fine distinction between military and police jurisdictions.

By the time the IRA accepted British conditions for a ceasefire in 1994, a source in the organization was prepared to admit that its ranks were seriously depleted. A large proportion of members of ASUs were either dead or in prison. Although guerilla cadres with significant expertise, in particular in explosives, remained in place, they were mainly limited to a 'core of old professionals', possibly totalling 'no more than fifty shooters and bombers in all of greater Belfast'. Dwindling support for armed struggle in Catholic communities, 'particularly among women', was discouraging fresh recruits.

The occasional coffee-jar bomb would still land on top of a British patrol. But the IRA's infrastructure was under such tight surveillance that most of its major operations were being pre-empted. 'A lot of information is leaking out through touts,' says a British Army source. 'The IRA's financial difficulties are making it easier for our special units to recruit informants.'

While the Army took most of the responsibility for counter-terrorism in Britain, other governments assigned the task to national paramilitary police organizations such as the French Gendarmerie, which had traditionally possessed special powers of arrest and intervention both in metropolitan France and its overseas departments.

After the GIGN's intervention in Djibouti, the SAS developed a close relationship with the French counter-terrorist team. Mike Rose, a fluent French speaker, frequently visited Christian Prouteau, and like two swordsmen the pair made it a ritual to go into the shooting range and compare their marksmanship. Prouteau tended to do better with a revolver and impressed Rose by shooting a clay pigeon off the chest of one his Gendarmes at a distance of fifteen metres. Rose, in turn, tried to lecture Prouteau on the British sense of humour. 'It is very important,' Prouteau remembers Rose telling him, 'for us not to take our jobs too seriously.'

The SAS became the first counter-terrorist unit to build a 'killing house', a building specifically designed for the realistic indoor practice of close quarter battle and room-clearing with pistols, sub-machine-guns and light explosives. At the time of their intervention in Djibouti, the only such facility available to the GIGN was an improvised target range which the Gendarmes had built by hand in an excavation pit outside their barracks. The GIGN was repeatedly invited by the SAS to use their killing house in Hereford. Eventually the French group built their own.

The SAS began developing a mini arms industry geared to the peculiarities of fighting terrorism, a 'low-intensity' war in which psychological factors count as much if not more than purely military ones. Their unique magnesium-based XFS stun grenades, or 'flashbangs', give off a blinding flash equivalent to 50,000 candles and a noise fifty times as loud as a backfiring motor cycle. The flash bang was designed to temporarily blind and deafen, and the worst injury it can cause is superficial burns, because it is packed in a cardboard carton whose metal mechanism drops off the moment it is in the air. It is the ideal 'softener' to save the lives of hostages while disabling terrorists during those critical few seconds in which a counter-terrorist assault force can access a room and seize the initiative.

On 13 October 1977 it was the turn of Germany's newly formed Grenzschutzgruppe 9 (GSG9) of the Federal Border Police to face a living nightmare. A Boeing 737 flying from Mallorca to Frankfurt had been hijacked with seventy-nine passengers on board, including eleven suntanned German beauty queens returning from a prize holiday in Spain. Commandeering the plane was a joint cell of four Palestinian and German terrorists demanding the release of imprisoned leaders and a cash ransom of $15 million. Two of the hijackers, including a woman, were members of the German Red Army Faction, which had assassinated the mayor of West Berlin, kidnapped the Chairman of the Board of the Daimler Benz Corporation, bombed NATO installations and now wanted back their two founders, Andreas Baader and Ulrike Meinhof, who were in jail. The other two were Arabs of the Popular Front for the Liberation of Palestine (PFLP), which had made aircraft hijacking a speciality. They had recently blown up a TWA airliner which they had forced to land in the Jordanian desert – just for show.

A group of GSG9 officers led by their charismatic commander, General Ulrich Wegner, came to the SAS 'Pagoda' seeking help. The Lufthansa plane had just landed in the Gulf state and former British protectorate of Dubai, and the Germans wondered if the British might have any useful connections there to facilitate a rescue operation. As it happened, an SAS officer was training the country's elite Royal Guard. The chief of the Pagoda, Major Alistair Moron, and his right-hand man, Sergeant Barry Davies, volunteered to join the GSG9 team and bring along a box of their flashbangs. The hijacked airliner took off from Dubai, however, before a rescue operation could be mounted, and went on to land in the People's Democratic Republic of South Yemen, which had kicked out the British a decade earlier and where terrorists now trained.

Negotiations with the terrorists were started in earnest as the requisitioned civilian jet carrying the special forces team circled around Yemeni airspace. The cash ransom had been agreed and the German government was prepared to fly the $15 million in cash to the terrorists, who chose the airport of Mogadishu in Somalia as their next stop.

After the terrorists unceremoniously dumped the body of the Lufthansa pilot who had been shot in Aden on the tarmac of Mogadishu airport, pressing their demands for the liberation of imprisoned comrades, the counter-terrorist team began preparing their assault.

Assisted by local airport workers, who lit a fire on the runway as night fell in order to distract the terrorists, the twenty GSG9 men and two SAS officers silently crept up beneath the wings of the airliner, then climbed over them using rubber-coated aluminium ladders to muffle any sound. Having set frame charges containing small amounts of plastic explosive, they simultaneously blew open both emergency escape doors. Major Morton and his sergeant were the first in, throwing their flashbangs, which rocked the cramped interior of the airliner with their intensely sharp, deafening explosions, blinding everyone inside like a ray of lightning, or as if 1000 flashbulbs had lit up simultaneously. Led by the SAS men, the GSG9 poured through the aisles firing bursts from lightweight MP-5 sub-machine-guns carefully aimed above the heads of the hostages, killing two terrorists and seriously wounding a third.

The final shots in the encounter came about seven minutes after the entry, when most of the passengers had been safely evacuated and the two SAS men were engaged in the enviable task of comforting the German beauty queens, who kept clinging onto them in various states of undress. A PFLP leader of the operation, Ahmud, who was near the cockpit when the assault began, had managed to hide in the toilet near the front of the aircraft. Whether he was hoping to evade capture and make a getaway, or planned to detonate a hand-grenade and take others with him in a desperate suicide, will never be known because when the GSG9 noticed the lavatory door was locked, they took no chances. One officer emptied the entire magazine of his MP-5, spraying it through the aluminium door at close range, then threw open the latch to find the Arab's bullet-riddled body sprawled on the toilet.

'What are flashbangs?' General Bernard Rodgers asked Charlie Beckwith the moment the colonel walked into the US Army Chief of Staff's office. News about Mogadishu had sent phones buzzing around the Pentagon, where the Joint Chiefs of Staff had received a personal note from President Carter asking if the American military had a counter-terrorist capability. Not only was the answer embarrassingly low-key but the top brass of the omnipotent US military establishment, which could annihilate the world in a split second, did not know about stun grenades.

Beckwith had survived Vietnam despite taking a .50-calibre bullet in his groin when his helicopter came under anti-aircraft fire as he flew to rescue his Delta Force from annihilation at the hands of a division of North Vietnamese troops. He came close to dying but returned to Vietnam 'for a gut check'. He had continued pushing for the creation of an elite counter-terrorist unit and happened to be in the Pentagon on the day of the Mogadishu rescue, experiencing the same stonewalling and bewildering maze of paperwork as Aaron Bank and other special forces pioneers who had preceded him. At five o'clock that afternoon he was walking to his car when an officer he recognized came running up to him. 'They want you up at the JCS,' said his friend, panting for breath. Word had quickly got round that if anyone had the answers for the questions which the White House wanted answered, it was Beckwith.

He was going through his recitation about flashbangs in front of General Rodgers when they showed him the note signed by President Carter. With the White House on the side of counter-terrorism, he could see the green light flashing. The type of unit Beckwith was proposing now had priority.

Otherwise, the Pentagon in 1977 was hardly the place to initiate any new unit, especially if it had anything to do with special forces. The Defense Department was reeling from budget cuts following the Vietnam War and the Green Berets were stigmatized with America's defeat in South-East Asia even if they had been the first troops to be pulled out when President Nixon began 'Vietnamizing' the war in 1970. By the time the North Vietnamese routed the South Vietnamese Army in 1973, there were no operational special forces units left in Vietnam.

Plans were underway to disband entire special forces groups and an increasingly active peace lobby, which had a strong voice in the new Carter administration was introducing legislation through Congress to prohibit American military personnel from assisting governments fighting internal 'popular struggles'. During hearings on the subject one influential Senator had flatly said, 'We are out of the business of counter-insurgency.' But guerrilla warfare in its menacing new incarnation as 'terrorism' – defined as high-profile acts of violence committed against non-combatants to further political ends – had become one of those universal evils, like crime or poverty, which everyone had to agree to fight. Even a relatively dovish administration like Carter's recognized that it was only a matter of time before a Munich, a Mogadishu or a Djibouti was visited upon Americans, and the new chief National Security Advisor, Zbigniew Brzezinski, had insisted at White House meetings that the USA needed to have the capability to respond effectively.

In November 1977 Beckwith got the $4 million, troop spaces and buildings he needed for his new Special Forces Operational Detachment – Delta, for he had chosen to retain the name of his former unit in Vietnam. He moved his headquarters into the old Stockade, or Army prison, at Fort Bragg, from where many special forces units had been recruited in the past. The thick walls and maximum-security cells were ideal for storing the high explosives, specially powered munitions, exotic weaponry and costly equipment which Delta was in the process of acquiring. With a blue awning and a rose garden decorating its entrance, the Stockade took on a certain distinction from the drab cement bungalows, World War II wooden barracks, nondescript depots, warehouses and offices which extended for many acres around the military metropolis.

North Carolina's Fort Bragg had become the Los Angeles of US Army bases, the place for soldiers seeking stardom through the action and glamour of airborne and special operations; the heartbeat of decision-making and innovation for rapid intervention forces. The Boy Scout camp atmosphere of Aaron Bank's day was being replaced by urban sprawl. Acres of pine forest had come down to make way for a jungle of communications and radar antennae, shopping malls, motels, fast food outlets and lots of low-budget housing. In the midst of all the bustle, Colonel Beckwith was like the upstart entrepreneur.

With personal access to the top echelons of the Pentagon, Beckwith could ruthlessly bypass the jealousies of other officers, including generals, along the regular chain of command who tried to stand in his way. He once cried in front of the Army Chief of Staff, telling him that there was a conspiracy to keep men in other units from joining Delta. 'Delta was the perfect place for Charlie,' says Bo Gritz, who at the time was commanding the 7th Special Forces Group based in Panama, 'because it was hidden. The man had no tact at all.'

'Charlie could say the stupidest things,' says Colonel Bob Mountel, commander of the 5th Special Forces Group and Beckwith's archrival, for he trained a team called 'Blue Light' to fill in the counterterrorist role during the two years which Beckwith projected it would take to form an instant SAS. Practising close quarter battle in an improvised killing house made of rubber tyres, Blue Light, Beckwith feared, was drawing resources away from Delta. Mountel had the seniority and manpower to seriously compete. But General Sam Wilson, whose career now extended from World War II Marauder to CIA field officer to Director of the John F. Kennedy Special Warfare Center to special operations planner in Vietnam, and who had recently turned down the post of Ambassador to Moscow to head the Defense Intelligence Agency (DIA), sided with Charlie. 'Blue Light was like a bunch of mechanics putting together a makeshift car. Delta, on the other hand, was a new design where engineers had to get involved with blueprints. It was a new Generation.'

Delta's model was, of course, British. As his second in command, Beckwith immediately appointed Major Louis 'Buckshot' Burruws, who had spent a year on attachment to the SAS in Hereford and passed their selection. He had come back with a note signed by all the NCOs saying that 'if you don't want this bloke we'll take him'. He spoke with a gentle Virginia accent and had rugged good looks. Beckwith and Burruws searched around the American south-east to find a place that most resembled Wales's Brecon Beacons as a site for Delta's selection course. They settled on the Appalachian mountains of West Virginia, where the only difference is that it snows more than in Wales.

Rangers, Green Berets, paratroopers and men from other units came in batches of 79, 185, 230 and 350 to try out for the role of James Bond through the physical and mental torture of a selection process which duplicated SAS methods in every way. Distances to be covered were the same: up to seventy-four kilometres; weights carried were the same: 55lb; the navigational work was the same and the questions in the psychological testing phase, to probe into the minds and attitudes of the candidates, were the same. The attrition rate was also the same. Those who passed trickled in: seven, twelve, twenty at a time, sometimes none. In a year Delta had about eighty men with whom to form its first squadron, which, as in the SAS, was subdivided into troops of about twenty men each.

Behind the Stockade, Beckwith built a killing house which cost nearly $1 million. 'I copied it from the SAS but made it bigger.' It was even equipped with the mock interior of an airliner. Lofty Wiseman, who was now an SAS demolitions expert with extensive experience in Northern Ireland was brought over to teach Delta the finer points of booby-traps. The SAS also helped Delta draw up a detailed checklist of all the factors that would have to be assessed in planning a mission to resolve a hostage crisis. How many people are on board the aircraft or building? What type and model of aircraft is being held? Everything down to the amount of carry-on luggage should be known. What are the physical characteristics of the pilot, the crew? Who are the passengers? Also needed was as much information as possible about the terrorist group.

Close liaison was established with other government agencies – the CIA, FBI, Secret Service. The Nuclear Emergency Search Team (NEST), a secret unit in the Department of Energy, worked out scenarios with Delta on how they would interface in the event of the ultimate nightmare: a terrorist group threatening a major city or installation with a nuclear device. Apparently, such threats had already been received from Palestinian groups. Delta's job was so sensitive and classified that no official record of the unit was entered on any government computer or filing system. Only a form filled out in pencil was kept locked away in a safe for purposes of processing pay cheques and other personnel requirements.

When the time came for an early evaluation exercise before the Army's readiness command, Beckwith was upset because the plane they provided was an old propeller model when Delta was already practising on a jetliner facilitated by the Federal Aeronautics Administration. Delta was also required to approach the target through a swamp. 'We are so far above this kind of shit,' ranted Charlie. 'This is Ranger stuff.' Delta just 'smoked the exercise'. Its snipers were so well concealed around a building where hostages were supposed to be held that someone playing a terrorist accidentally urinated on one of them. The assault then came with 'total speed and violence'. Swinging through a window like Tarzan, one Delta man landed directly on top of a 'terrorist' sitting on a bed. Another role-player trying to block a door went into sheer panic when a Delta sledgehammer came crashing through the wood just inches above his head.

Exactly two years after its activation, just on schedule, Beckwith declared Delta to be 'operational' and arranged for a more serious evaluation exercise. This time the counter-hijacking was done on a Boeing 727, which had to be 'taken down' simultaneously with a building where hostages were also held. Watching the show were Sam Wilson, General Peter de la Billière, who then headed all British special forces, GSG9's General Wegner and GIGN's Christian Prouteau.

With the lowliest rank of anyone there, Captain Prouteau confesses to having felt somewhat embarrassed and surprised when standardized questionnaires were passed around asking all the observers to grade the various phases of the exercise. 'They asked us, how did you like the sniping? Mark the appropriate space: excellent, good, average . . . How did you like the counter-hijacking? etc.' Prouteau and others considered it the wrong approach in evaluating special forces, one which showed a certain lack of understanding on the commander's part. 'Delta was trained to an extremely high level,' he told Beckwith, 'but you don't give them the confidence to achieve. In Europe we would never grade them like this.'

De la Billière commented that since the decision to intervene in a terrorist situation was usually a political one, either President Carter or some senior member of his administration should have also participated in the exercise. 'In Britain when we play, Maggie [Thatcher] also plays.' But when the time came for Beckwith to play with Jimmy Carter, there was no role acting. It was for real.

The day after Delta's debut, as Charlie recovered from drinking Jack Daniel's all night at the big party thrown by Sam Wilson at his motel, he got the news: the US Embassy in Iran had been taken over by an armed mob and fifty-four State Department officers were being held hostage.

By 1980 hostage takings were bursting out everywhere and when the SAS got the call, Lieutenant Colonel Rose at least didn't have to wrestle with the problems of inserting a force into a hostile country halfway around the world whipped up into a frenzy of xenophobic hysteria. Instead, 22 SAS staged its counter-terrorist extravaganza in London's Embassy quarter, overlooking Hyde Park, in what is referred to in CT talk as a 'permissive environment'. At 11.48 a.m. on 30 April 1980, B Squadron was being rotated through CRW training, practising its daily routine in the killing house, when six Arab terrorists armed with sub-machine-guns, pistols and grenades burst into the Iranian Embassy in Princes Gate. 'A flashbang reverberated in the room as I fired double taps neatly into three terrorist cardboard dummies when the alarm sounded,' recalls Soldier I.

B Squadron packed their equipment into holdall bags, got into a convoy of unmarked cars and vans and drove from their Hereford base to London. By the time they arrived at an Army barracks off

Regent's Park at midnight, Michael Rose was already at his command post on the sixth floor of the Shell-BP building, overlooking the besieged embassy. The recently appointed SAS commander was, at that stage, 'military adviser' to the Cabinet Office Briefing Room, or COBRA, the crisis management team chaired by Home Secretary William Whitelaw which now gathered to deal with the situation. The Cabinet ministers and security officials represented in the special committee had their different roles well rehearsed. The military option was a tool of last resort — the final solution once all other efforts to resolve the siege peacefully were exhausted.

The Mujahidin Al Nasser Martyr Group of the Democratic Revolutionary Front for the Liberation of Arabistan was threatening to blow up the Iranian Embassy with all the hostages unless demands were met for the diplomatic recognition of an independent Republic of Arabistan in southern Iran and ninety-one of its members were released from Iranian jails. Recruited and trained in Iraq, the terrorist group had received arms and final instructions in London from a military officer working under diplomatic cover at the Iraqi Embassy. The seizure of the Iranian Embassy amounted to a special operation orchestrated by Iraqi dictator Saddam Hussein to pave the way for a military invasion of Iran's vital oil-producing region. He had a plan to spark an insurrection in Arabistan.

'The terrorist leader, Salim, was a highly intelligent and educated man,' according to Rose, 'and his group were well armed, highly motivated and trained. We took their warning that they had the embassy booby-trapped seriously. For us, it was a closed situation.' As negotiations dragged on through the second day, Rose ordered B Squadron to move up to a holding area next door to the embassy. They arrived as night fell in rented Avis trucks, making their way through back gardens into the state-owned premises of the Royal College of Nursing.

As the deadline which Salim had set for the following afternoon passed without incident, police negotiators managed to talk him into releasing a BBC sound recordist suffering from a heart condition who was among several British journalists seeking visas to Iran and a police constable who had been trapped in the besieged building. In return for his cooperation, the terrorist was given assurances that Arab Ambassadors would be called in to mediate and his demands would be broadcast on the BBC World Service.

Salim was particularly interested in the publicity and threatened to kill a hostage when the BBC edited his statement. 'You are not dealing with my demands. You are sitting on your fat bottoms doing nothing. I tell you now. Because of Britain's deceit, you British people will be the last to be released. And if you do not send back the BBC man to talk to me, someone will have to die.' The police negotiator promised Salim that Scotland Yard's Director of Information would personally read out his statement word for word over the evening news on condition that he release two more hostages.

'It had been decided between us that if two hostages were killed, the SAS would go in,' says Rose who had an 'immediate action plan' to take the embassy if the slaughter started. 'The original plan was to run out of our holding area to the embassy next door and batter down the ground-floor windows with sledge hammers,' says Soldier I. Rose preferred a margin of two instead of one dead so as to be sure that multiple killings were taking place before launching a raid in which he calculated a probability of 90 per cent casualties among the hostages. Information available, at first, on the inside of the embassy was very sketchy. The SAS had no idea of where in the building the hostages were being held. Those who had been released reported that they kept getting moved around. B Squadron was alerted and stood down several times on that tense day.

Soldier I would periodically check his German Heckler & Koch MP-5 sub-machine-gun, a weapon the SAS had adopted since the GSG9 had tested it in action at Mogadishu. It had been found better suited to close quarter battle, firing straighter and more accurately, than the American Ingram or Israeli Uzi which had been the standard sub-machine-guns until then.

Rose was simultaneously working on a 'deliberate action plan' for a more studied assault and the key man to help was soon located. An ex-officer in the SAS had been the consultant in charge of designing the security for the Iranian Embassy. He kept detailed plans of the building which allowed the SAS to construct a full-scale plywood replica in an Army facility in London where B Squadron went to rehearse their assault room by room. He immediately told Rose that the ground-floor windows and main door were armour-plated. Sledgehammers would have bounced off. Demolitions equipment was now issued, mainly in the form of frame charges: threads of plastic explosive, usually of no more than a few grams, which are taped to windows and doors to blast them out. It also became possible to determine certain patterns of activity inside the building. 'Two terrorists were awake at all times,' Mike Rose found out from one of the released captives; 'and the hostages were segregated in different rooms'.

As electronic intelligence specialists began placing listening devices through the walls of adjoining buildings, Soldier I and a team from B Squadron went for a night-time recce of the embassy's roof. Snipers were already positioned at the Royal College of Nursing, Rose's headquarters in the Shell-BP building and among the bushes in Hyde Park. Through the forest of telecommunications apparatus, aerials and satellite dishes, Soldier I located the embassy's skylight, managing to lift it open after fifteen minutes of stripping off the lead waterproofing around the glass. He looked into a cramped and dirty bathroom. By the moonlight, he could make out a door which led into the rest of the embassy. 'I felt a sudden rush of excitement. A surge of adrenalin at the thought of becoming the first SAS man to enter the embassy.'

At this point the SAS might have opted for a 'silent entry' in which a small assault team armed with silenced weapons would have stealthily crept into the building to kill the terrorists one by one and free the captives. But the ongoing negotiations and continuing lack of intelligence on the whereabouts of the hostages ruled out such an option in favour of the developing plan for a 'distractive entry'. In this preferred scenario the building would be taken from all sides, simultaneously, by a large force using a maximum of noise and violence, when the circumstances warranted. While the skylight facilitated an entry point into the top floors, an interior route into the ground floor was being engineered by removing the bricks separating the embassy from the Royal College of Nursing, leaving only the plaster between both buildings, which could easily be blown off with frame charges. At one point Salim noticed a bulge which this made in the wall. Trevor Lock, the police constable who had been taken hostage near the start of the siege, assured him that it was caused by a leaking drainpipe.

'We had gained the initiative and there was hardly any stress at all in planning the deliberate assault plan very methodically,' says Rose. 'Salim's group had taken on a stronghold which was too big for them to control.' As the terrorists became worn down by the endless negotiations studiously stretched out over tedious details by the police, 'their tactics became less well put together'. The terrorist operation had originally been planned for forty-eight hours. The siege had gone on for twice that time. Feeling increasingly confident with his plan, Rose walked over from his command post to the holding area on the evening of the fifth day to deliver his final briefing. The men of B Squadron sat in folding chairs, wearing their full assault kit and holding their respirators before the scale model of the embassy as Lieutenant Colonel Rose breezed in.

The assault force would be divided in two teams. Red team would handle the top floors; Blue would take the bottom half of the embassy. Rose took them step by step over how eight men from Red would abseil down from the roof in two waves onto the second-floor balcony and break in through the windows. Another group from Red would descend through the skylight and by ladder onto the top floors. Meanwhile Blue would break into the ground and first floors from the back, over an adjoining balcony and through the wall. All the finer points were covered, down to the removal of dead bodies. From the released hostages the SAS had also gained detailed physical descriptions of all the terrorists and the hostages, which they now memorized from police identikit sketches and a few photographs.

Final preparations for the deliberate assault were being completed just as the siege went into its sixth day and Salim was reaching the end of his patience. His last proposal to be driven to Heathrow with the hostages to board a plane for a friendly Arab nation, à la Munich, had gone unanswered.

Feeling boxed in, restless and ultra-paranoid, Salim woke up at 6.30 a.m. and told PC Lock that he had been hearing strange noises all night. The terrorist leader may well have heard sounds from the roof where SAS men were preparing their abseiling equipment. But his megalomania was clearly getting the better of him when he asked Lock to check the building and report back if any of his police colleagues had entered. He really believed that Lock was on his side.

'Your police are up to something, I am convinced,' said Salim. 'I'm going to make new arrangements for the hostages.' His gunmen pulled their checked *shemags* over their faces and herded all the males into the telex room on the second floor, where they remained guarded by the two most fanatical of the terrorists, armed with a sub-machine-gun.

Pleading with Salim for his permission, Lock and one of the other British hostages, Sim Harris, ran over to a window and shouted to the police negotiator outside, 'Time is running out, where are the Arab Ambassadors who are supposed to mediate?'

'Things are moving along as quickly as possible,' came the reply. 'The Foreign Office are still in discussions with the Ambassadors and if you hear the BBC World Service you will get your confirmation.' The 1 p.m. news bulletin, saying that no decision had yet come out of the discussions between the Ambassadors and COBRA, only added to Salim's anxiety. With the expressionless face of a man suppressing his worst fears, he picked up the phone, dialled the police duty negotiator and said in a drab monotone, 'There will be no more talking. Bring the Ambassador to the phone or I will kill a hostage in forty-five minutes.'

When bursts of automatic fire were heard coming from inside the embassy, Soldier I instinctively reached for his MP-5. 'I removed the magazine, cocked the action and caught the ejected nine milly round. Then I stripped the weapon down and began cleaning the working parts meticulously. This is it, I thought. There could be no going back now. A hostage had been murdered.'

There was a second burst of sub-machine-gun fire as an urgent meeting of COBRA debated whether one or more hostages had really been killed. No dead body had been presented. Knowing the fanaticism of Arab guerrillas, for he had fought in Oman and Aden, Rose was the only member of the committee who felt certain that a killing had taken place. 'The timing and conditions seemed about right for the killing to start; for Salim, a psychological barrier had been crossed.' He informed COBRA that pending some last-minute adjustments, his deliberate assault plan 'would be ready to launch on five minutes' notice'. Home Secretary Whitelaw considered it prudent to continue the negotiations. The SAS advised him that the assault should be launched while it was still daylight.

'We put the body on the doorstep. You come and collect it. You have forty-five minutes. Then I give you another.' Salim's flat message came over the telephone into the crisis control room at 7 p.m. Rose turned to his TV monitor to see the crumpled body of the Iranian press attaché pushed out onto the embassy steps and come rolling down. Prime Minister Margaret Thatcher was watching the same scene on her television at 10 Downing Street. After consulting with Whitelaw and having a telephone discussion with Rose she gave her consent to launch the assault, reminding everyone that the whole world was watching and that success was 'absolutely necessary'. The final projection which Rose gave COBRA was still 60 per cent casualties among the hostages. The Metropolitan Police Commissioner went through the formality of scribbling a brief note on a sheet of lined paper, passing control of the operation over to Lieutenant Colonel Rose.

'I inserted the clip with thirty rounds into the curved magazine of my MP-5. There was the usual reassuring metallic click as I snapped home the cocking handle and applied the safety-catch.' A torch was bolted onto Soldier I's stubby sub-machine-gun for his particular job of clearing the building's dark basement. He rose from his cot and fitted on his body armour, fastening the side straps of the Kevlar layers covering his chest and torso. He then slipped his assault vest, carrying the flashbangs, over the armour. A 9mm pistol with a thirteen-round clip of ammunition was inside the holster strapped to his thigh.

He looked around at the other men now hastening through the last-minute preparations necessary before assaulting a terrorist stronghold. They would be shooting their way through fifty rooms containing hidden gunmen and, as far as they knew, booby-traps, with the objective of locating and rescuing nineteen hostages. Their last-minute intelligence brief could only allow an educated guess as to where the hostages were being kept. Despite their endless room-clearing drills at the killing house, it was the first time that the SAS was doing it for real. Something once written by Paddy Mayne, the war-loving Ulsterman who co-founded the regiment with David Stirling, flashed through Soldier I's mind: 'When you enter a room full of enemy, kill the first one that moves – he has started to think and is therefore dangerous.'

He glanced at Sek, the tall black Fijian, now fitting on his A6 respirator. Eight years earlier they had fought together in Oman's battle of Mirbat, hopelessly outnumbered by hundreds of Adoo guerrillas armed with Chinese weaponry and launching human-wave attacks. From the roof of the SAS post, Soldier I had fired his fixed Browning .50 machine-gun, which became so hot that it gave him blisters, while Sek breech-loaded 25lb shells into an old field gun, shooting it at point-blank range until, wounded by several enemy rounds, he could no longer move. Now he recognized several other members of his squadron who had been among the relief force which had helicoptered in to save them when they had been practically overrun. It was infinitely reassuring to know that they were all behind him once again.

At Mirbat they had fought almost naked, wearing sandals and shorts. Now the same men stood covered up monstrously in their black fire-retardant suits and plated armour, hooded under their respirators with built-in radio headsets. They appeared like high-tech recreations of some medieval knightly order, the men of the 'flaming sword' or 'winged excalibur', uniquely tried for the ritual combat of counter-terrorism. When Blue team's leader, 'Hector the Corrector', pulled the respirator down over his disfigured faced, permanently scared by an Adoo bullet, the rest followed suit.

Twenty minutes after depositing the first dead body on the embassy steps, Salim was taking a call from the police negotiator, who asked, 'Shall we park the coach at the front of the embassy?'

The terrorists now heard lots of noises, becoming ever more distinct. Glass smashed as a Red team abseiler got snagged in his ropes and hit a third-floor window trying to untangle himself. Soldier I, running up the back garden, saw the man swinging above just in time to stop the detonation of the frame charge on the french windows. 'It's all going wrong, I thought. There's no way we can blow that charge without injuring the bloke.' They had to resort to using their sledgehammers, making more unplanned noises, which, as things turned out, saved the day.

'Get back!' other members of Blue team had to shout to Sim Harris, who was gesticulating at them from behind the reinforced front windows, to save him from being cut to shreds by their demolitions work. Alerted by the sound, a terrorist, pointing his sub-machine-gun, appeared on the second-floor balcony – and right into the telescopic cross-hairs of an SAS sniper concealed across the street in Hyde Park.

'We were quite lucky,' admits Rose. 'Our sniper engaged the terrorist guarding the hostages who had the only sub-machine-gun.' The Czech-made Skorpion dropped uselessly onto the ground below the balcony when the terrorist was jolted by a head shot that blew his brains out. The first bullet fired by the SAS at Princes Gate assured the success of the operation – fortunately, in view of everything else that was going wrong.

Explosive charges and flashbangs rocked the building. Part of the ceiling collapsed, setting fire to the top floors and scattering debris and broken glass everywhere. The gathering blaze cut access into the building for the part of Red team going through the skylight, where overcharged explosives had caused much of the damage. Three abseilers had made it onto the rear balcony of the critical hostage-filled second floor. But they had to worry about their team leader, who was still tangled in his ropes above them, swinging dangerously close to the burning curtains of a third-floor window.

He was finally cut loose from his harness by the second wave of abseilers, but took a hard ten-foot fall onto the balcony which added serious bruises to the pain of his severe burns. Hurling in their flashbangs, the eight men stormed through the windows only to find nothing in the general office where they had been briefed to expect hostages. They realized that instead they were barricaded inside by furniture pushed up against the doors from the outside corridor. A critical minute had been lost.

At his command post in the Shell-BP building, Rose took a moment to watch Kate Adie standing before the gutted embassy, which was now reverberating with explosions, gunfire and screams. 'It seems that things aren't going too well in there,' the veteran BBC trouble-spot reporter commented in her usual understated manner. 'Three minutes into the thing, it did look extremely bad,' says Rose. But the SAS commander was satisfied from the radio reports he was receiving from his team leaders wading into the chaos that 'we had managed to hit the most important floors simultaneously with a lot of people. As long as that was kept up, I was confident that we would succeed.'

Fire was sweeping through the first floor, where hostage Sim Harris had been rescued, running into the safety of the adjoining balcony as SAS troopers poured in through the curtains set ablaze by flashbangs. 'Trevor, leave off!' came the muffled command of an SAS corporal shouting through his respirator when PC Lock was found wrestling with Salim in a corner office. The policeman had finally produced the revolver kept hidden in his tunic for the past six days but was not quite able to fire. He was pushed aside as Salim was drilled in the head and chest with fifteen MP-5 rounds at point-blank range. At the opposite end of the floor, the torch on another MP-5 cut through the smoke and tear-gas hissing out of CS canisters fired into the Ambassador's suite. A terrorist was caught in its light just as he tried to take cover behind a large sofa. He waved his pistol, receiving twenty-one rounds in return.

Meanwhile confusion prevailed on the second floor, where Red team finally broke out of the general office through a corner door and into the small cipher room. Screams went up from the four panic-stricken women hostages huddled inside the cubicle at the terrifying spectacle of the black-hooded figures appearing from out of the smoke and debris. One member of the team, Trooper Tommy, a stocky Scot, had on his own initiative climbed back out of the general office and onto the window ledge of the adjoining filing room. Crashing through the glass, he caught sight of a terrorist trying to set fire to the files. He aimed his MP-5 and pulled the trigger, but it jammed. Then his eyes watered and he coughed from smoke seeping in through his damaged respirator. He pulled it off, revealing his shock of red hair. Throwing the mask aside with the sub-machine-gun, he drew out his pistol and went after the terrorist escaping into the corridor.

Shots rang out from the telex room at other end of the second floor, where a terrorist armed with a pistol was shooting the male hostages. If the sub-machine-gunner guarding them had not been taken out by the SAS sniper, the massacre might have been complete by now, possibly surpassing Rose's projected 60 per cent casualties. As it was, two hostages had been killed by the time two other terrorists ran inside, flung their weapons down and tried to blend in among the terrified captives bunched up against the far wall. But the fanatical guard kept firing, wounding another hostage before Tommy kicked open the door.

Crouched in his CQB instinctive firing position, the SAS man swung around in a ninety-degree arc, his pistol gripped firmly with both hands, and took aim at the guard, who now produced a grenade. Before the terrorist could pull the pin, Tommy squeezed off one 9mm bullet, drilling a hole in the terrorist's forehead, killing him instantly. A moment later three more black-hooded figures appeared through the door. They had to move the hostages out fast before fire consumed the floors below. There was not much time left to do a thorough check of the terrified captives.

'I was now conscious of the sweat. It was stinging my eyes and the rubber on the inside of my respirator was slimy. My mouth was dry and I could feel the blood pulsing through my temples.' Soldier I had cleared the cellar, finding nothing. Only for a moment did he think he saw a terrorist, and emptied his MP-5 into what turned out to be an empty rubbish bin. The adrenalin-fuelled elation which he had felt just seconds before was starting to dissipate and as he ran back up the basement stairs, 'the body armour which had felt as light as a T-shirt when I led the way in now began to weigh on me'. He and Sek moved through the library, which was not ablaze thanks to the dangling leader of Red team, whose plight had stopped them from blowing their plastic charges through this entry point. The moment in which he had laid eyes on all the books, Soldier I had also held back from throwing his flashbangs. Otherwise they would now be trapped in a blazing inferno.

'The hostages are coming. Feed them out through the back, I repeat, out through the back,' came the orders over the headset from Major J., the commander of B Squadron. As he approached the main staircase, Soldier I could see the masked figures of his other team members lining up on the stairs. He took his place six or seven steps above the hallway, letting the MP-5 hang on its sling around his neck so as to free his hands to help the hostages and point them in the direction of the back garden. The distressed women stumbled down first, shocked and confused.

'This one's a terrorist!' The high-pitched warning filled the reception area. Soldier I focused on a dark face ringed by an Afro haircut, above a body bent over unnaturally. 'Then I saw it – a Russian fragmentation grenade. I could see the detonator cap protruding from his hand. Adrenalin rushed back in. I slipped the safety-catch to automatic on my MP-5. But, Shit! I can't fire.' Other soldiers were standing in his line of sight. 'I instinctively raised the sub-machine-gun above my head and in one swift, sharp movement brought the stock of the weapon down hard on his neck.' The man's head snapped back, the tortured, hateful face looking at Soldier I as he collapsed, rolled down the remaining steps and two sub-machine-guns were emptied into him with a deafening echo.

'In that split second, my mind was so crystal-clear with adrenalin that I zoomed straight on the grenade pin and lever. I stared at it for what seemed like an eternity and what I saw flooded the very core of me with relief. The pin was still located in the lever. It was going to be all right.'

The commander's orders crackled through the headsets again. 'You must abandon the building. The other floors are ablaze. Make your way out through the library entrance at the rear. The embassy is clear, repeat, the embassy is clear.' The last of the terrorists hiding among the hostages was arrested in the backyard as the SAS body-searched all of them. Ali Abdullah was found to have 9mm ammunition in his pockets but was otherwise unarmed. They had to take him alive.

The men of B Squadron thirstily cracked open the cans of cold lager being handed to them as they sauntered back into the Royal College of Nursing for the victory celebration. While they laid aside their weapons, body armour and belt kit, Margaret Thatcher suddenly appeared for a surprise visit to congratulate her prize special forces team. A flushed and smiling Michael Rose strolled in beside her. 'There is nothing sweeter than success,' she told her cheering, sweaty group of champions. 'And you boys have got it.' She went on to speak about 'brave and brilliant management . . . faultless teamwork . . . and immense physical courage'.

But more than any other factor, it was the individual initiative characteristic of SAS soldiers which carried the day at Princes Gate. No one had to order the sniper hiding in Hyde Park to shoot the sub-machine-gunner who appeared on the second-floor balcony. He hit the critical target the very instant the opportunity presented itself. No officer instructed Trooper Tommy to crawl over the window ledge into the filing room. It was not in the original plan. He did it on his own initiative and proceeded to rescue most of the hostages despite his failing equipment.

Under pressure Soldier I made a series of correct decisions which proved critical not so much because of what he did but because of what he didn't do. In refraining from detonating explosives and his flashbangs at the library entrance to the ground floor, he prevented causing a further fire which could have blocked the main exit route and seriously jeopardized the final stages of the rescue. Despite his frustration at not having scored a kill that day, he held back from firing his sub-machine-gun at the terrorist coming down the stairs when he saw the risk of hitting his own mates. 'Commanding special forces is the easiest thing in the world,' Michael Rose says, 'because once you give them the objective you can rely on the fact that they all know what they are doing.' 'When you burst into a terrorist stronghold every man has to be his own troop commander,' says an instructor from 22 SAS's sister service, the Australian SAS.

5

GEMS IN THE ROUGH

In a remote corner of Western Australia, at the foot of the Stirling mountain ranges, Private Stewart Bailey sits cross-legged inside a pup tent, barefoot in torn fatigues and a T-shirt, longish hair falling over one eye. He is busily cleaning his rifle after crossing sixty kilometres of thorn-bush country. Weighed down by a 55lb rucksack, Bailey has struggled over three mountain peaks in temperatures of 100°F in three days. He is the first in his course to successfully complete the hardest phase of SAS selection, 'Happy Wanderer', and I am there to witness it.

Nobody is pushing the men, who have been dropped alone into the wilderness to prove that they have the stamina, endurance, iron will, and most important of all, superhuman motivation to be accepted in this most elite of special forces units. They can pull themselves out at any time and be returned to their regular units. 'It's like extracting the pure gem from under the earth,' is how the process of selection is described by the Australian SAS commander, Lieutenant Colonel Higgins, 'which we then cut into a fine diamond.'

The tall, lanky farm boy from Adelaide exudes a sense of self-satisfaction betrayed by a reserved smile as he oils and cleans his weapon. Its bolt, coil, magazine and other working parts are carefully laid out on top of a sheet in front of him. He says that he is a loner – the very quality that the SAS looks for. In what is soon to become his former unit, the Royal Australian Regiment, the twenty-one-year-old Bailey has already proved himself in his specialization of sniper.

He attaches a bronze brush to the point of a rod and pushes it into the rifle barrel as he speaks. 'My father made money in farming and the spring-water business and sent me to agricultural school. But I had always wanted to go into the Army and joined the infantry when I was seventeen. I enjoy soldiering and never want to sit behind a desk.' He sees the SAS as a natural progression from his current role in his battalion's reconnaissance platoon. 'I will get to go to different countries and work with a lot of different armies, specially the Brits and the Yanks.' Bailey expects to join a Mounted Vehicle Troop to maraud behind enemy lines in jeep patrols like 22 SAS did in Iraq. He soaks a little oil into a piece of flannelette which he rubs gently over the chamber.

'My ideal kind of reconnaissance sniping patrol would consist of just myself with a rifle and maybe another man with an automatic weapon in support. We would be out for several days to get within range of an enemy position and take out personnel and equipment. By hitting the officers and radio operators I could render a company-sized unit useless. But not just men – radios, thermal optic sighting in tanks, even artillery, can be rendered ineffective by effective sniping. . . From a distance of 400 metres I can score accurate head shots.' He snaps the bolt back into the rifle's cleaned-out chamber, raises the gun and looks through its sights.

'The difference between the SAS and other units is that while regular soldiers may find themselves fighting on their own if things go wrong, an SAS man has to be prepared to operate alone at all times,' explains a former commander of Britain's 22 SAS. 'That is his normal mission.' During the Falklands War, a four-man patrol inserted to set up an observation position near the main Argentine base of Port Stanley operated for twenty-eight days totally cut off from any reinforcements or resupply. Having gone in with rations for only a few days, they functioned for sixteen days trying to fix the position of enemy helicopters before a resupply of food could be delivered to them. A few extra grams of chocolate or biscuit would be apportioned each day to the man whose turn it was to trek some twenty kilometres with a heavy radio to relay their reports back to headquarters so as not to risk detection by enemy radio intercepts. They lived in 'scrapes', holes dug into the wet ground, covered by soaking peat in freezing rain and snow.

The Argentines moved their helicopters every day to avoid their location by the British, usually landing them in well-concealed valleys. During the month-long game of hide-and-seek, the four SAS men managed to fix the position of the helicopters on three occasions. Each time it was reported, Harrier jump-jets took off from the aircraft carrier *Hermes* for an air strike and twice the raids had to be aborted because the helicopters were moved on time. On the third try, the SAS patrol finally guided in a successful air strike which destroyed most of the helicopters, denying the Argentines a critical advantage which they may have otherwise held over the British.

'There is no other unit in the world which could have accomplished that,' says 'Bucky' Burruws, deputy commander of the US Army's Delta Force, who went through selection with the SAS. 'Any other soldiers I know would have just quit. Selection is the only way to make sure that you get the type who won't.'

'I cant explain why selection works,' says John Woodhouse who instituted it at the SAS. 'It just does.' Charles Beckwith, who imported it for Delta, describes it as 'pushing yourself beyond any imaginable limits of your mental and physical endurance and having the guts to reach for that extra reserve, that indefinable quality which keeps you going further.'

'They will take you to the extremes of your physical endurance and get five more miles out of you,' says Sergeant Mick McKintyre of 22 SAS. 'The question is how you can overcome your own physical exhaustion by your mental ability. To continue when every sinew of your fibre is screaming stop.'

Under the supervision of John Woodhouse, the Australian SAS adopted the British system of selection when its first squadron was being trained for the Borneo campaign. 'The Australian and British SAS are mirror images of each other,' says Sergeant Steve Paterson, whose Mounted Vehicle Troop practises annually with 22 SAS in the deserts of Oman. Both units sport the sand-coloured beret with the insignia of the flaming dagger and scrolled motto 'Who Dares Wins'. The British 4x4 Land Rovers are petrol-powered while the Australian six-wheelers run on diesel and pack somewhat less firepower, lacking the mounted 44mm cannon which the British vehicles carried in Iraq. But otherwise, the combat doctrine practised by the *shemag*-covered warriors of both units is taken from the same book – figuratively speaking, for the SAS has no published field manual.

While the main physical danger at the 22 SAS selection site in Wales's Brecon Beacons is hypothermia, in Western Australia it is dehydration and the midsummer 110° heat knocks out half of the eighty-six soldiers who step off an air-conditioned C-130 transport plane at the military airstrip of Bindoon. They are immediately ordered to run up and down the melting tarmac in their full combat gear for the equivalent of 3.2 kilometres or until they drop from sunstroke-induced exhaustion. Even Stewart Bailey falls apart, later confessing that it was the moment in which 'I came closest to pulling the pin' – the SAS euphemism for quitting. Forty-three of the men voluntarily drop out during the 'pre-selection' phase which also includes, after the run, one hundred sit-ups, twenty heaves and fifty dorsal raises. 'Since they've all failed, we leave it up to them,' explains Sergeant Craig Goss, a supervising NCO.

'Not bad,' shouts Goss two days later at the tired-looking men laid out before him in a pile of green and brown camouflage. 'Most of you achieved two checkpoints a day; not all of you did two a night. Only one got to all. Do any of you have any questions on how to use a map?' The reply is muted. They have all just completed a navigational march over flat and featureless pasture land to make sure that each has the ability to use a map and compass. Getting lost up in the Stirling Range could prove much more hazardous. 'All any of you want to do is board the truck back home, isn't it?' Goss challenges them, with a knowing look.

Two trucks back up and they all pile in, cradling their rifles and lying against rucksacks which feel like blocks of steel. Deep salt marks outline the sweat stains on their uniforms as a cloud of stench rises inside the covered truck. They compare their scores: most reached seven of the eight checkpoints, but a few got to only five. There are comments about that 'Kiwi instructor', the native New Zealander who gave them a particularly hard time. All look forward to a shower, tea and a night's rest after their two days trekking through the flat, monotonous countryside.

They are in for a what is known to the Australian and British SAS alike as a 'sickener'. The men have barely had a chance to lie down on their beds when they hear 'All right, out you go!' and they are plunged into a punishing session of PT. Half are ordered to do sit-ups while the rest box. The two groups get switched around. 'Hit harder. I wouldn't mind fucking your missus if you was coming home angry,' screams the instructor. They go into a frenzy of pounding as punch-bags get knocked to the ground. 'It's called controlled aggression,' quips the instructor.

'Vaughan, run some laps!' he barks. Vaughan has a dislocated shoulder which prevents him from punching. But he can still walk, run and carry a rucksack. As long as he can keep that up, he remains fit for the course. Not able to alternate with the boxing, he has by now done 300 sit-ups. The running is a welcome break and they will all soon be running several kilometres into the night before being finally allowed to lie down.

'It's an aggressive masculine existence,' says Gregg Jack, the head of the SAS training squadron, as he strides into the Sergeants' Mess to join his other NCOs for his nightly session of beer drinking. Jack has at various times been attached as an adviser to 22 SAS in Hereford and Delta Force at Fort Bragg, North Carolina. Some of the sergeants settle down to watch pornographic videos while can after can of beer is cracked open. It's out of place to be without one. The group gets rowdier, the jokes get dirtier and the woman on the screen is now performing fellatio. Every other word is 'fucking' as the NCOs compete in hurling insults at the 'politicians' who held them back from joining the British SAS and Delta Force in western Iraq during the Gulf War.

'Prime Minister Bob Hawke wanted to send us,' insists Porky, 'but the other bunch of arseholes in his cabinet stopped him.' Jack's eyes actually water when he says, 'We missed our Guernsey.' The humorous Dusty adds, 'We're like the koala bears, mate, we're a protected species.' Then they all break into the chorus, 'SAS stands for stay at Swanbourne' – the Perth beach-side suburb which houses their base.

Porky has left the room and returns with a handful of artillery projectile simulators, ready for another sickener. The SAS selection candidates have been sleeping for only an hour when he walks into the darkness, pulls the pins from the projectiles and hurls them at the tents. The slow whistle of falling artillery shells breaks the stillness of the night and is followed by ear-splitting bangs and pink flashes. Cries and moans rise from the tents as the men crawl out with their rifles and form up in three rows. They look dazed, and are shaking. Some can barely stand up, aching for the sleep of which they are being deprived. Porky orders two to do twenty push-ups for not having their bush hats. Another one has to do fifty for missing his webbing, the vital belt and suspenders arrangement which carries ammunition, water and emergency rations and is as much a part of any soldier on duty as his limbs. Some of the men haven't put their trousers on – but that's all right.

The officers have been on the move, with only three hours' sleep, for almost two days and they now have to hike six more miles through the dry bush country to uncover an arms cache and prepare an ambush on a Long Range Patrol Vehicle (LRPV). They are deprived of more sleep than the men in order to test their ability to plan and make decisions under conditions of extreme fatigue. 'We try to identify those who can maintain a clear and logical thought process when the rest of the body is functioning at 50 per cent alertness.'

Crawling through a bush of prickly vines, the six officers identify the LRVP parked on a road. Observing the target from different angles, they sit down and each sketches out an attack strategy. They then break into a three-kilometre run to a rendezvous point where trucks and a hot meal are waiting. Lieutenant Bridgeford is suffering from a dislocated ligament in his right leg, and is sweating profusely. He calls to the others, 'Are you all right, fellows?' then picks up the gallon tank of water that is being passed between them and keeps running.

When he arrives at the RV point, Lieutenant Larkey's feet are badly blistered, with bandages covering both of his swollen soles. A medic unwraps his feet, bursts the blisters with a syringe, draws out the puss, carves out the hardened skin, injects antiseptic and applies a layer of synthetic skin. It is in this condition that Larkey boards the truck which takes him and the others to the Stirling mountains.

Next time I catch sight of Bridgeford, he is reaching the top of Mount Toolbrunup, which at 4000 metres is the tallest peak in Western Australia. It took me an entire afternoon to climb up there through steep bush-covered tracks, slopes of flat rock and gigantic boulders. With a bergen not even half the weight of Bridgeford's, I was exhausted with creeping dehydration, having drained both my water bottles long before getting to the top. The sun is starting to set, and Bridgeford still has a torn leg ligament and just can't talk. Only the intensity of his obdurate expression betrays the extreme fatigue and pain he is experiencing.

After refilling his canteens from a jerrycan, Bridgeford pulls out the heavy radio from his pack and reports the checkpoint code 'Tango Oscar' back to exercise control, giving his call sign. Sergeant Brett Rowland, the supervising instructor at Tango Oscar, hands the lieutenant his next objective – Hotel Echo. Bridgeford sits down to look it up in his notebook, works out the grid reference on his map and the route he is going to take to Henton Peak. It is 14.6 kilometres from his present position and he will have to negotiate a way through other mountains to get there. He reports the information to exercise control and is off, hoping to snatch two hours of sleep on the way.

Those on selection retain little sense of time. Every man just has to keep moving with an unrelenting momentum and has a maximum of five days in which to complete three mountain peaks. Days turn into nights and nights turn into days. All normal needs are out of their minds. 'You don't want to sicken yourself out,' says one. According to Rowland, Bridgeford is only the second candidate to turn up at Tango Oscar on that first day of Happy Wanderer.

The star-covered sky above the Stirlings gives the clearest view of the universe the naked eye is likely to get from earth. Constellations, shooting stars and man-made satellites are all visible as an ever-present reminder of the transient nature of life in the eternal battle-field, filling even the most drained body and mind with a sense of infinity. Sergeant Rowland sits beneath torchlight, reading a book on satellite reconnaissance and military strategy. But for the forty-three SAS candidates trekking alone through the desolate bush, the nature of their military mission becomes very fundamental. Exhaustion, pain, hunger and thirst are its only manifestations as they become reduced to little more than struggling organisms straining to break through the limits of their physical beings. 'While hacking my way through the scrub,' says one soldier, 'all that I could see in front me were prickly vines. It took me an hour just to advance 500 metres in the darkness. The weight I had on my back kept making me fall backwards. Suddenly, it was as if I had lost my sanity. I just wanted to cry, bawl my eyes out and scream for mother. Then, just as suddenly, something snapped. I felt all right again and just kept going.' Those breaking through their bodily barriers are also being prepared for their special role on this planet, that of facing death unflinchingly and disposing of life without hesitation.

I wake up in the middle of a cloud the next morning. It's cold and I can barely see a foot in front of me. 'You all right, mate?' Rowland's face, covered by a black woollen balaclava, materializes out of the fog, offering me a mug of hot coffee – the best I've ever had. It's 8.30 a.m. and the fog is still thick when Private Bishop appears on the peak. He is in a state of adrenalin-pumped euphoria and can't stop talking. 'It feels great to know that I've done the highest peak. All the others will be small compared to this.' He sits, lays down his rifle, unslings his rucksack and begins to take his bearings with his map and compass, pointing to a mountain past a track in the far distance. 'That's my next one.' Magog Peak is nine miles away. He energeti-cally repacks his radio after reporting his route to base, straps his bergen back on and picks up his rifle. 'See you later, mate,' he says cheerfully as he disappears through a gap in the rocks. One of the qualities which the SAS men get marked on, along with stamina, endurance, initiative, mental guile and others is a sense of humour in adversity.

From the top of Toolbrunup, I can see patrol vehicles below acting as 'enemy' to keep SAS candidates off roads and tracks. All of the men's movement has to be through the bush. As the sun begins melting the clouds, a soldier from the Training Squadron turns up on the peak, lugging two gallon cans of water. The helicopters normally used to resupply the checkpoints are grounded because of broken parts, he explains. Water and food can only be carried up by hand, limiting the quantity which can be delivered. The men on selection are now faced with the added hardship of half-rationed water. The soldier has instructions to escort me down the mountain and soon we are on our way, negotiating our way through the steep boulders around the peak. Until the helicopters can get airborne, he explains, no effective search and rescue can be mounted.

Shots suddenly ring out. The soldier called John who barely scraped through selection himself two years ago and was admitted into the Regiment's support unit, moves quickly and sees that one SAS candidate is stuck on the ledge of a steep crevice. He has run out of water and feels dehydrated. His bergen is too heavy for him to make the climb up the rocks and he is going into a panic, uncertain even of where he is. John gives him a drink from his water bottle, gets on top of the rock to pull up the bergen for him and orients him towards the checkpoint. Realizing that he is close enough, the SAS candidate regains his composure and decides to keep going.

Further down the mountain, Corporal Paul Dumbavin stands looking though his magnified prismatic compass to fix his position relative to a hill just west of him. The feature is marked on his map and by drawing an imaginary line in accordance with the degrees marked on his compass he is able to determine that he is halfway up Toolbrunup. He rests briefly, leaning on his rifle. 'My father fought with the SAS during the Malaya Emergency,' he explains. 'This is a family tradition.' Whistling a tune, he resumes his march up the slope.

Back at the base camp radios are crackling at Gregg Jack's Exercise Control tent as the chief instructor charts each candidate's progress on a blackboard with the reports coming in from the different checkpoints. Bridgeford has TO141847 and HE150800 before his name, meaning that he arrived at Hotel Echo from Tango Oscar at

8.00 a.m. on 15 February. Those who have dropped out have a blank line drawn from their last checkpoint. Eight have quit so far and Jack does not expect many more. 'Most drop-outs are in the first couple of days. They usually decide to pull the pin during the first night when they realize that they don't like being out in the bush alone. Anyone who has made it this far, on the other hand, is likely to continue. The types we are looking for will be experiencing a second wind and thriving on it, relieved that they are on their own without a thousand eyes bearing down on them like at Bindoon.' Most pull-outs are marked WOR (Withdrawn by Own Request). Three are for medical reasons: injury or ligament strain.

Others could be getting lost. 'If we don't hear from one of them for a couple of days, a rescue mission is sent out.' Saying this reminds Jack to turn to one of his aides and tell him, 'Check to see if those bloody helicopters can get up yet.' The two American-made Royal Australian Air Force Blackhawks are parked in a clearing just outside the tent. Jack sprints out to work off the tension with some heaves on a thick branch of a tree.

One candidate comes wandering into the base camp. His crazed blue eyes seem to shoot out of his sockets as he mouths off incomprehensible phrases like some shell-shocked soldier who has staggered out of a desperate battle. 'I don't see the point,' we finally understand him saying. 'Just don't see the point.' A sergeant walks over and leads him into the medical tent. He is suffering severe sunstroke. The number of drop-outs by the end of day two is nine.

On the afternoon of day three, Vaughan, still suffering from a dislocated shoulder, is crawling to the top of Mount Trio, croaking, 'Give me a drink, just give me a drink, I've got to have a drink of water, mate . . .' The Army rugby player has gone without water for fifteen hours. The instructor insists on getting his map grid references sorted out first, but Vaughan rolls on the ground like a stricken dog. 'A drink of water, mate, I've got to have a drink.' He has just finished a four-kilometre climb in the middle of the day. Although he has considered pulling the pin all along the bush-whack through the tangled vines, he later confesses that his crazed thirst rendered him incapable of making a rational decision. The sergeant finally hands him a canteen, which Vaughan empties into his mouth. He decides to keep going.

Lieutenant P. has gone without water all night and throughout the morning. It is now noon and the lashing heat is squeezing out every last drop of moisture from his exhausted body. He collapses for a moment, panting, trying to get what relief he can from the little shade beneath the bushes. There is only one thing to do, he decides. Standing up, he unbuttons his flies, takes the green plastic water bottle from its canvas container on his belt, holds it beneath his penis and urinates into it. After throwing in a water-purifying tablet, he drinks the mixture.

The next morning Private Carnie arrives. For him it's the fourth peak and although his face is twisted up with pain he talks a bit. 'I've preferred this part of selection in which I'm on my own, can call my own paces and lay down to sleep whenever I want,' he says, almost repeating the words of the chief instructor back at Exercise Control. 'My motivation will never leave me. I come from a parachute regiment which is fucking boring. The routine just gets to me. I hate the unit I'm in.' As he straps his bergen back on, Carnie reveals the bottom line: 'I've got nothing to go back to.'

The Blackhawks are finally airborne on the fourth morning and they can be seen circling at a distance above the various peaks. The deafening whop-whop of the rotor blades followed by the whirl-wind of their powerful down-draught breaks the stillness of the Stirlings as they come hovering over Tango Alpha to rope down supplies of food and water as these are running critically low. Before getting on his radio to communicate with the pilots, Dusty pulls down his jogging shorts and sticks out his backside to greet them with the 'brown eye'.

I didn't meet the star of Happy Wanderer until I got back to the base camp at noon. Stewart Bailey had made it back early that morning, hiking all night from his last checkpoint. He had already showered and seemed rested when I entered his tent. His score sheet read, 'well above average performance'. In addition to making his three checkpoints in three days – something of a record for the SAS – during pre-selection he accomplished a two-mile run with a 55lb bergen in less than sixteen minutes; a twenty-kilometre forced march with full equipment within the required space of three hours with thirty minutes to spare; and a swimming test of 400 metres, twenty-five metres of which were below water.

It was hard at first to get him to talk in anything other than monosyllabic phrases. 'Yeah . . . it's all right . . . pretty hot, I suppose . . . feel good . . . a bit tired . . . found thorny bushes more of a problem than mountain climbing.' He demonstrates another characteristic of the SAS man: to say little about experiences which are beyond the reach, capacity and even understanding of most normal men.

After leaving Bailey, who was clearly born to join the SAS, I see a group of soldiers sitting in a circle in front of another tent. Like Bailey they are isolated from the rest of the base camp and I notice how they look away as the star private passes by holding his mess kit on his way to chow. Introducing myself, I realize that they are officers: Lieutenants Steven Pata, David Hanzl, Michael Mahy and Angus Donald. They all have another thing in common – they have failed selection and are waiting to be RTUd, returned to their units. For some reason, the five enlisted men who dropped out were sent off immediately while the four officers are being made to wait. There is no available transport, apparently.

Pata quit on the second day. 'When you are hoofing it alone, all by yourself, you get time to think. I started to examine more how I felt rather than how I was actually performing and I started to miss my wife and children. I realized that this just wasn't for me . . . I look forward to going back to my post in the Royal Corps of Transport. It's a good job.'

The other three lieutenants also have families or are recently married, whereas Bailey and the other men I encountered making all their checkpoints, it dawns on me, are bachelors. 'If I had done the course two years ago, I would have continued,' says Hanzl, who comes from an airborne regiment and had wanted to join the SAS for eight years. 'But now that I'm married my priorities have changed. My wife didn't want me to join the SAS as it meant being away from home too much. She just started meaning more to me.' Mahy admits, 'I just couldn't take the pain of my bloody blistered feet,' adding, 'I also have a month-old marriage to worry about.'

The US Navy SEALs, who have maintained a close relationship with the Australian SAS since Vietnam, with frequent joint exercises and personnel exchanges, advise their trainees to remain isolated from their families during the weeks preceding 'Hell Week'. The SEAL selection process is considered among the most exacting and physically strenuous training programmes of any of the special forces. Ron Yaw, who led SEAL hit teams during Vietnam's Phoenix Program and rose to command its elite counter-terrorist unit, maintains that, 'Nobody who can pass a standard psychological test is likely to make it. It takes reactions to things which are not normal. You've got to have a warped attitude.'

Hell Week is a five-day ordeal of constant physical demands and contrived hardships with practically no sleep. While alternating between running obstacle courses, crawling through sewage, swimming for miles in cold sea and paddling boats for more than eight hours at a time, the men are never allowed more than one hour of intermittent rest. 'After the first two days, you are in a total mental fog,' says Yaw. 'You just do what's next. Capacity to think has gone away. Some other part of your brain is controlling your responses. I don't know where it is, maybe in your ear lobe.'

Unlike SAS candidates, SEALs are kept under close and constant supervision during Hell Week. This is partly for their own safety. Serious injuries, infections, pneumonia and severe mental disorders will occur. It's not unusual for the men to start hallucinating in the middle of a swim or pass out while paddling through a heavy surf. But deliberate mental harassment is simultaneously applied by the instructors, whom Yaw describes as 'very warped psychiatrists. They can detect what an individual's mental weakness is and attack it to try and make you quit.'

During one three-hour run along a misty beach, Yaw had an instructor driving in a jeep beside him. 'Tired now, Mr Yaw . . . Why don't you just give me that stupid red helmet, get in the jeep and I'll take you home . . . Why are you doing this to yourself? Nobody even wants you here . . . you could be in your own warm cabin on a nice ship with a good hot cup of coffee . . . Doesn't that make more sense than trying to be something you are not . . . ? Come on, Mr Yaw, jump in. No one wants you in the SEALs, don't you understand?'

'Your mind just locks up,' explains Yaw. 'It freezes and blocks out everything except: he can't make me quit. It becomes an all-encompassing total thought, blocking out physical pain and any emotion or discomfort. You loose any conception of past, future and present. Nothing the instructor says has any real meaning. That's what keeps you going. The times when you are most vulnerable are when the pressure is off. That's when you really begin asking yourself, why am I doing this? It's best during those moments not to even know what's coming next.'

If a man falls asleep, the instructor pours water over him. He will deflate the rubber boat which each one has to carry at all times and force him to reflate it manually. Then there is 'The Circus' – rigorous callisthenics after the daily twenty-one-kilometre run, swim and paddle which go on until somebody drops out. They want it to get to the point where it isn't just the instructor but the candidates themselves who want each other to quit to relieve their torture. The 'nobody really wants you here' feeling becomes real.

The 'no quit' attitude is perhaps more important among SEALs than other special forces units because their prime environment is water. On land, a soldier overcome with exhaustion can usually hide somewhere and just rest on the ground. A SEAL cannot do that in the sea, which is not only his primary means of insertion into a target area but is often his only escape route. During the American invasion of the island of Grenada in 1983, a SEAL platoon was ambushed by a Cuban-led force equipped with armoured vehicles. Despite receiving several hits, including a bullet in his elbow, the team leader, Lieutenant Erskine, led his men back into the water, and they swam for several miles out into the Caribbean until they were picked up by a destroyer. SEALs have to look for safety to the water, where sleeping means drowning.

Being comfortable in the water is a prerequisite of entering a naval special warfare unit, a role which in Australia is filled by the SAS. Britain's Special Boat Section (SBS) of the Royal Marines, which at various times has run its selection courses with Australians and US Navy SEALs as supervisors, submerges its candidates in a flooded gravel pit. Breathing air in through a tube they must stay under for up to half an hour to test their ability to withstand the claustrophobia induced by being in dark, murky water. A majority of volunteers drop out during this initial phase of pre-selection.

Ron Yaw, who was a champion swimmer in college, was one of only three men left from the initial course of eighteen who started the day of prolonged submersion testing. Most quit during the diving locker phase in which they had to walk down a ladder into thirty feet of water in metal diving suits and remain submerged for one hour. Four pulled out instantly. In the final phase, four men went down together in a recompression chamber to a depth of sixty feet, with the pressure inside being gradually increased to three atmospheres. Yaw ended up making the fourth descent on his own.

In a stress test at a more advanced stage of SEAL training, the men are required to jump into a swimming pool with their hands and feet tied. A diver underneath feeds them oxygen from his tank as they sink into the water. It is just another way of making sure that the SEAL is totally comfortable in the water and is capable of a totally abnormal action.

'Special forces across the world share a different mental framework, a psychological disruption which allows them to get through training,' says Ron Yaw, who has had extensive experience working with elite counter-terrorist units from various countries as commander of SEAL Team Six. 'It's a glue which has kept special forces the way they are. It's facing a very dangerous situation and conquering it, overcoming the maximum adversity which gives you the ultimate thrill.'

Such a mind-set, however, can prove counter-productive. During the Grenada invasion, a squad of four SEALs tried parachuting 600 feet into a rough sea with winds of over twenty knots, each carrying an equipment load of over 100lb. They drowned as they got tangled in their parachutes, being dragged down into the water by the excessive weight. The equipment could have been dropped separately in pallets for the team's rendezvous with boats which were supposed to insert the men on to a beach, but the SEALs felt that they had to prove that they could carry it. The aborted SEAL reconnaissance mission seriously jeopardized an Army Delta operation being conducted in Grenada the next day. 'The SEALs are somewhat more gung-ho,' says a former commander of the SBS.

France's GIGN, which also trains for highly hazardous maritime counter-terrorist operations, considers it vital to test the 'reflective' capacity of its recruits. 'It's very important to have a highly balanced person,' says Captain Denis de Favier, 'one who is not easily scared but who thinks as well.' During a rigorous, week-long selection course in which GIGN assesses an average of one hundred candidates for ten places, they have to run eight kilometres with a 50lb load, make a vertical climb of ten metres on a rope, swim 100 metres with their hands and feet tied and do fifty metres underwater. But a key part of the test is observing the reaction of a man when he is told to jump from a twenty-five-metre tower harnessed to a rope. 'If he refuses, he is naturally out. But if he does it too quickly without the moment's pause in which to reflect, he is out as well. What we want to see him do is check the security of his rappelling equipment, make sure that his harness and belt are firmly strapped on, that the rope is properly attached, and then jump.' The French are more philosophical.

* * *

'All right, push!' Red faces turn purple, heaves become screams and the men pushing on either end of the iron pole fixed to the ram bumper of the six-wheel-drive Long Range Patrol Vehicle bend all the way over, exerting every iota of pressure they can squeeze out of their drained bodies. You are reminded of ancient slaves dragging immense building blocks to build pyramids. The LRPV which the team of twelve SAS candidates are pushing up a hill moves just a few inches before they all collapse.

'Your effort is average!' yells out Sergeant Joe Van-Doroffelaar, the legendary 'Jungle Ghost' of Vietnam, who still serves in the Australian SAS training squadron. He speaks with a thick stutter, adding, 'Three of you are about to throw in the towel.'

'I don't really know what they want,' gasps one of the thirty-four suffering survivors from the Stirling Range who have now been plunged into the biggest 'sickener' of all, the three-day 'Lucky Dip'. The final phase of selection is called that because among the things they have to do is cross a river, which comes to most of them as a very welcome relief. Having thought that they had finished after their eighty-kilometre mountain hike, the SAS candidates are now required to perform the most tedious and physically gruelling tasks in their burned-out condition. During Lucky Dip they are also subjected to close and often degrading scrutiny by the instructors. The truck is not going to get moved up the hill. The sergeants standing around watching the men struggling all know that. But the NCOs who manage the SAS want to get an even closer look at the thirty-four finalists, and whittle them down further as the pressure gets turned up to super-stress and beyond.

Van-Doroffelaar pops a sweet into the mouth of his hard-bitten face. He belongs to the oldest and most experienced generation of the Australian SAS, the Regiment's inner core of NCOs who fought in Vietnam, and worries that, 'We have been at peace for so long that we are no longer sure of what our standards really are. This is the only way of testing the men's calibre, by making them pick fly shit out of pepper with boxing gloves on.'

In the Stirlings you suffer and go down or triumph. In Lucky Dip you just suffer as selection begins to take on the aspects of a surreal nightmare, a warped, upside-down world in which failure can be the only result. The men are starting to look and act like zombies and feel like them, if they are lucky. 'My legs hurt at the beginning,' says one. 'You can imagine what it's like now. I've got a migraine up to my chest, just numb from pain.'

'My body is drained,' says another. 'My mind is still alert but my body has difficulty moving. I feel like I'm going in slow motion, the same as when you get up in the morning. I'm in a daze.'

Sergeant Mick McKintyre remembers his Lucky Dip while on selection with 22 SAS during winter in the Brecon Beacons. He was required to cross a frozen river, breaking through the three-inch ice with his rifle butt as he went along so that freezing water came up to his chest. 'Up to my legs, I had no feeling. I didn't feel the freezing effect. The body process just slowed down as if your batteries were being drained. My blood was thickening and slowing me down as if I was trying to run in waist-deep syrup.'

Van-Doroffelaar is scrutinizing the best officer material. He points to Lieutenant Ben Larkey, whose feet were bandaged up with blisters before he even got to the Stirlings. 'He will make a good troop commander. He knows how to keep command and control.' The two other lieutenants on the team, Van-Doroffelaar thinks, 'are no good'. Larkey urges on his flagging team with patient encouragement as they try pushing the LRPV up the hill once again. He keeps his finger on the pulse at all times, switching the men around to get the most out of the strongest, whom he concentrates along the front and on the back of the vehicle. One soldier who barely manages to keep going with codeine injections for his injured back is made to push along the sides. Larkey, who is the son of a civil servant, makes up in leadership abilities what he lacks in manual skills, having admitted earlier, while struggling to fashion a stretcher out of a piece of canvas and two poles, 'I just can't tie knots.'

A burst of automatic gunfire rattles the patrol moving through the bush, carrying tank tracks to supply the fictitious guerrilla army of U Gum Shit. The cry of 'Ambush!' goes out as all twelve men dive for cover, forming a protective perimeter and a burly sergeant appears out of the bushes holding up his rifle. 'We are moving tactical,' he says, bearing down on the SAS aspirants, 'and you are turning into a bunch of rabble. Patrol leader is too far out in front. You are not covering your arcs of fire . . .' He replaces the lieutenant leading the patrol with a private.

In the SAS the NCOs pick their leaders, having as much if not more say than the commissioned ranks on which officers get accepted. This uniquely egalitarian system has led to charges from certain quarters that 'the pigs rule the sty'. But while officers get rotated out of the Regiment every three years, the non-commissioned ranks are normally expected to remain with the unit for the duration of their military career. Enlisted men receive a considerably higher pay increase in the SAS than officers, who are often referred to as 'Ruperts' and whose function is often perceived as little more than administrative. During the two main phases of selection, Happy Wanderer and Lucky Dip, officers are treated on an equal footing with enlisted ranks – something which happens nowhere else in the military – even as their ability to inspire leadership is closely measured.

Approaching fifty, Joe Van-Doroffelaar has experienced almost the entire history of the Australian SAS. But his mean, scarred look contrasts, somehow, with the more clean-cut Anglo-Irish appearance of most of the other soldiers around him. One could imagine him, the son of a Dutch submarine captain who defected to Australia with his vessel when the Japanese overran Indonesia, fitting in better with the Foreign Legion. When I tell him this, he looks away with a sardonic smile, takes a pensive drag on his cigarette and dismisses the comment. 'Naa, Foreign Legion's no good.' Then he agrees to sit and talk to me.

He believes that the Australian SAS jungle and bush patrolling tactics are still the best in the world. 'As good as they were in Vietnam and better than any other white man's.' The British, he thinks, 'are losing it because they've been in Northern Ireland too long,' although 'I wouldn't recommend Arctic warfare for the Aussies.'

Complaining that 'there is a growing tendency for special forces to rely too much on technology', Van-Doroffelaar insists that, 'It cannot be allowed to overtake the five senses. Such aids as Night Vision Goggles can enhance a mission and although they have been improved to give better depth of vision at night, they still tend to cut off lateral vision, which is essential for a man on patrol in order to detect enemy movement around him. A special operator should actually develop a sixth natural sense, a proximity sense.

'It's only a matter of time before we do less training in manual navigation and begin relying on Global Positioning Systems' – portable computerized satellite-guided maps which give a man his exact position complete with grid reference and distance relative to a programmed target area. The GPS is now in common use by 22 SAS and the USA's Delta Force and SEALs. 'If these gadgets are lost or break down, a soldier has to know how to fall back on himself,' adds Van-Doroffelaar. 'Technology greatly enhances the stand-off means for reconnaissance. That's very important. But in direct action missions, special forces will always need to get right up to a target. That is why we have to continue to place more emphasis on the man.'

'It's rats,' says an SAS candidate as he uncovers the pot of boiled guinea-pigs which have been brought out for the starving men. It is their first hot meal in a week and their first resupply of food. By now they have all consumed most of the six ration packs each was issued with at the Stirlings. 'I feel my body starting to bite into its reserves,' says one man. 'It's used up all the fat and is starting to eat up the protein, which is muscle.' Most are starting to look emaciated, having lost an average of 10kg each. Fit young men used to the suppleness of highly developed muscles now feel limp, actually sensing the nourishing effect of what they believe is rat meat going into their bodies pass into them as they tear off the skin and white pieces of flesh from the boiled animal. The broth from the guinea-pigs also relieves the coldness in their stomachs. None of them knows how much longer it will all go on for or when their next meal will be.

It's day four of Lucky Dip and a ten-man team is pushing a wheelbarrow improvised out of wooden poles and a tyre to carry ammunition boxes weighing several hundred pounds. It creaks along as the men constantly struggle to balance it and steady it and prevent the contraption from coming apart as they pull and push over the bumpy trail. Some are harnessed to it by rope like draught horses. Glazed expressions are fixed on every face. There is hardly a reaction when a small kangaroo or wallaby bounces across the track. Even Bailey is beginning to look run-down, like he has had a hard day on the farm. Yet on they go through mile after mile of monotonous bush country, past thickets of thin trees which give practically no shade, dry leaves crumbling between the heavy feet and hard ground, manoeuvring around giant ant hills which resemble primitive ruins.

Several days' growth of beard bristles from the men's soiled faces. Uniforms are torn and the deep salt stains of sweat disfigure the camouflage pattern. The stench can be sniffed for yards around. Eyes widen in the effort to remain awake. No one is going to drop out at this point. Most of the men are too tired to even think about it, reduced to the state of paraplegics accustomed to the aches and pain of basic movement. It's like watching a chain gang of condemned men being slowly driven to their death, and evokes Australia's origins as a penal colony.

The procession rolls along. The makeshift wheelbarrow's constant squeaking and screeching strains the nerves as the instructors walk beside it, their eyes intent on the degree of effort each man is exerting, scrutinizing every nuance which betrays an attitude. 'He's just plodding along, just not fucking working,' says the tall, square-built, hard-nosed Sergeant 'Kiwi', the native New Zealander who has become notorious for his toughness since the course began. He is talking about a Navy diver. 'He would be going straight into water operations, which means counter-terrorism, the most prestigious side of the SAS. See how he leans up against a tree. He would need constant supervision. When you break into a terrorist stronghold, every man has to be his own troop commander.'

Orders go out for the team leader to be switched and a freckled captain gets put in charge of the doomed command. Onwards, barely moving now as the wheelbarrow starts falling apart, heads go down, men start dropping off, one gets held up by his mates as his knees give in, the captain shouts out instructions, trying to get the group back together. 'Fucking useless,' exclaims Kiwi about the officer, 'he talks too bloody much. The troopers he would command would just walk all over him.' The officer is still rattling out orders when they reach a road and Kiwi is really getting annoyed. As he sits down to mark his score sheet he finally says, 'Eh! Will you bloody well shut up!' Just as abruptly as it started, selection is over.

Of the thirty-four men who came down from the Stirlings, twenty-three are finally selected for 'continuation training' with the SAS. 'Three of them are edgy,' I am told while eating my hamburger at a barbecue in the NCOs' mess at the beach-side base outside Perth. Both winners and losers have been invited. One soldier whom I met during Happy Wanderer walks into the party having just finished his meeting with the CO, Lieutenant Colonel Higgins. 'I'm shattered,' is all he can say. He didn't make it. Vaughan, who got through the entire ordeal with a dislocated shoulder and half-dead from thirst, walks in looking equally incredulous. Much to his surprise, the stocky, five-foot-seven semi-professional rugby player is now among the chosen ones. 'All that kept me going was this image of a pint of cold beer and that big steak I was going to have at Zizzler's.'

Some of Australia's Rambos might have stumbled over some of the questions they were asked in the final phase of psychological evaluation, such as, 'If you were on patrol behind enemy lines and were discovered by a young girl, what would you do?'

Captain Rod Boswell, a British Royal Marine, was doing quite well on his selection course. Twenty-two SAS apparently liked the way he took off his clothes before crossing a cold stream, carrying them on top of his head so that he could put them on dry when he got to the other side. 'You are the first one who ever thought of that,' an NCO told him admiringly. But Boswell came unstuck over the question of the little girl. They drove him hard: 'What if you knew that she was a double agent and the sister of the resistance leader you were working with?'

'The girl obviously needed to be silenced. But I could not do it myself or order someone else to do it. In the heat of battle, perhaps. But cold-bloodedly . . .' Boswell just couldn't come up with an answer which he felt was the correct one. Feeling that he was being indecisive, he voluntarily pulled out.

A paratrooper on Boswell's course, on the other hand, immediately said that she should be raped and killed and left on the road so that the enemy would think that she had been the victim of a sexual assault. He thought that was what the SAS wanted to hear. However, they sent him back to the parachute regiment, recommending that he be given a thorough psychiatric examination.

By not coming up with a definite answer, Boswell had in effect offered the correct one, he would later learn. It's Catch 22. The SAS are testing a man's reaction to being faced with a very difficult decision – whatever was done with the little girl would ultimately depend on the circumstances. During the Gulf War, SAS and American Special Forces patrols were faced with the dilemma in real life. Neither killed the little girl and as a result of being compromised three SAS men died in action and four were taken prisoner by the Iraqis. The eight-man American team had to fight their way through a battalion of Iraqis for an entire day before being pulled out by helicopter.

Delta Force uses a psychiatrist. 'If a psychiatric evaluation comes in saying that a man has just so much paranoia, that's good,' says Charlie Beckwith. But Delta's first squadron was well into continuation training when Beckwith became concerned about one of the men. He had done very well in the physical part of selection but kept getting carried away on the firing range, 'just hosing down those targets with a sub-machine-gun like there was no tomorrow'. He was sent back for further psychiatric testing and the revised diagnosis noted 'psychotic tendencies'. Beckwith returned the man to his unit, but in his opinion the SAS would have kept him.

Delta even screens the wives of its married officers to determine how supportive they will be of their husbands – a factor which is considered in the men's selection. Beckwith instituted this screening informally by arranging for the women to meet with the highly intelligent and intuitive mother of his deputy commander, Bucky Burruws. 'A very special and super lady' is how Beckwith describes Frances Burruws. At the age of eighty-six she was still working with Virginia's state correctional authorities in prisoner rehabilitation. Frances explains how she would get to know the wives of the prospective officers socially and discreetly gauge their attitudes. 'In one case, I remember how one woman seemed totally uninterested if not downright negative about her husband joining Delta.' I advised Charlie about this and he decided against taking him. 'He wasn't all that good, apparently, and if his wife was going to be a problem, Charlie figured that it wasn't going to work out.'

Beckwith also required Delta candidates to pass a strenuous swimming test 'to keep out them goddamn niggers'. Blacks tend to be bad swimmers and as an unrepentant racist, Beckwith believed that they tended to make bad, lazy soldiers and didn't want them in Delta. 'The problem with blacks,' says British SAS man Lofty Wiseman, who worked closely with Beckwith, 'is that their big muscles burn up too much oxygen. Their frame and large pores makes them poor swimmers because they get waterlogged.' But even qualified underwater divers sometimes fail the 500-metre swim in uniform and boots and have to retake it until they got used to moving in the water without fins.

In Australia, at least, there is a hedonistic side to the masochism of the SAS. Nude bathing is strictly protected on the beach in front of the Swanbourne barracks. As it is government property under SAS control, the local authorities cannot interfere. A panorama of naked breasts and golden pubic hairs generally greets the men on water exercises, infiltrating the shore in their motor-powered, eight-metre-long rigid inflatables. While conducting his own reconnaissance along the beaches, Sergeant X once came across Lulu, a thirtyish buxom blonde divorcee. 'He was beautiful to look at,' recalls a leggy brunette lying at the beach with Lulu that day. He sat down with them, chatted up Lulu in his humorous way, they played in the waves and he got her telephone number.

When he appeared to fetch her at her doorstep in a beaten up Land Rover, she opened the door wearing tight jeans and a midriff blouse showing everything up to the peaks of her Stirling Range. After a seafood dinner at a shoreside restaurant and a couple of bottles of the white wine from Western Australia's vineyards, they went back to her house and Sergeant X was feeling beneath her blouse around the most gorgeous breasts he had ever touched. Soon she melted and her jeans were unzipped. Sergeant X just lifted her up with his muscular arms and fitted her on to him as he walked towards the bedroom, carrying her up and down the house with her legs dangling around his hips. Through open doors, out to the garden, then back in again, he kept it up for what seemed like an eternity. Lulu's orgasmic cries went past the point of deep moans. In an expression of sweet total surrender, she just locked her legs around him to enjoy the ride of her life. It was something that she had never experienced with her former hotel executive husband, she later told her friends. The SAS will neither confirm nor deny.

6

CUTTING THE DIAMOND

A blast of cold air fills the inside of the C-130 transport aircraft as the ramp opens up into the night sky. Captain Boyd Parsons hooks his static line on to the anchor cable and looks at the green light which flashes on. The jumpmaster pushes out two equipment bundles and Parsons immediately follows them out, plunging into the dark vacuum. First he feels a slap from the propeller blast and, after a breathless moment, the relief of being yanked up by his inflated canopy. In the next few seconds he looks up to see the opening parachutes of the other eleven men in his A team.

The landing zone seems to be getting blacker as he drifts to the ground. Signal lights which marked the spot for the airplane are now extinguished. Only vague outlines of the valley and mountains are visible moments before his feet touch the ground and he goes into a well-practised roll to prevent the impact, equivalent to a fifteen-foot fall, from harming his spine or any vulnerable extremity.

As he unharnesses his chute, straining through the pitch-blackness to see if the rest of his men are making it down safely, Parsons hears the rustle of approaching footsteps. He drops the harness and releases the safety-catch of his M-16. There are men fanning out around him. Sparks and then the flame of a match illuminate the profile of the man he has come to meet. 'Robin Hood', wearing a fedora, is lighting a pipe which gives Parsons the call sign for his reciprocal move. From the pocket of his field jacket, he produces a pouch of Turkish tobacco. Robin Hood accepts it, checks the manufacturer's label and smiles. The bona fides have been established in the crucial 'G meeting' where the Green Berets make their initial contact with the guerrilla force they have been sent to support – a moment which can be a life or death affair.

For the next month Parsons and his team will be closely advising and leading Robin Hood's insurgent forces operating in the wilderness of the Uwharrie National Forest in North Carolina. It's a realistic war game conducted by the US Army Special Forces to test the Green Berets in the skills, physical endurance and mental stamina required to conduct guerrilla operations. Robin Hood's forces consist of men drafted from regular units, such as military engineers, who have no training in specialized tactics or survival skills. Furthermore, they possess little motivation other than personal pride and a taste for adventure to stand the constant discomforts of being cold, wet and hungry for a seemingly interminable period in the pine jungle.

'The challenge is to be able to lead and motivate others whose identity and reasons are very different from your own,' says Parsons over a pitcher of beer at the bar of the Green Beret Club in Fort Bragg. The walls are covered with billboard posters of all the movies which have been made about them. The one for *Green Beret* is personally autographed by John Wayne. Outside are the World War II grey wooden barracks which housed the first American Special Forces group, formed by Aaron Bank in the 1950s, and are now the headquarters of the Army's Psychological Warfare Brigade. The Green Berets and newer units like Delta Force have moved on to more modern facilities.

'There are three types of people who join Special Forces,' Parsons explains. 'Those who like to work secretly on their own, those who want to go to some exotic place and kill somebody and those who enjoy working with and getting to know different types of people.' The last time I had met the Green Beret colonel he was the chief US Special Forces liaison officer with the Arab armies during the Gulf War. 'The desert is our friend,' Parsons told me then. If he hadn't gone into the Army, the veteran Green Beret thinks he would have become an international sales representative for some big corporation. The large, square-shouldered special operative is now out of his 'chocolate-chip' desert wear and in his green jungle fatigues. which blend in better with the surrounding pine forests in which he first trained. 'I prefer this part of SF,' says Parsons. 'I would find a specialized counter-terrorist unit like Delta to be highly limiting.'

Only a few nights ago, I am told, military police came roaring into the area around the Green Beret Club. They had not come to break up one of the frequent punch-ups which erupt there but in response to explosions. Delta had been conducting one of its secret exercises in the vicinity of Smoke Bomb Hill. To keep anyone from observing their special tactics for taking down a bus, Delta had not informed the Fort Bragg authorities, and had cleared the area by the time the MPs arrived. In one case in which MPs did catch a group of seedy-looking men in civilian clothes and sub-machine-guns in the process of what seemed like a kidnapping, it took a great deal of explaining to convince them that it was another typical Delta night practice session.

During the exercise in the Uwharrie National Forest, Parsons leads his 'Merry Men' against an elite battalion of the 82nd Airborne Division, in which he previously served. The paratroopers are highly motivated to oppose the guerrillas, having been promised three to five days' leave for each Green Beret they capture. The insurgents are raiding facilities, ambushing convoys, gaining intelligence on the enemy and generally disrupting rearguard activity over a wide area in preparation for a conventional army assault. An underground network of local civilians assisting the guerrillas with food supplies, hiding places, and information operates throughout the farms and villages. But any of the corn-pipe-smoking hillbillies and friendly farmers' daughters could also be enemy informants.

While observing a highway bridge over a period of several days, Parsons contacts some local farmers to find out if there is any enemy reaction force in the vicinity. 'I ain't seen none,' he is consistently told and radios for 250lb of explosives to be sent up from a clandestine guerrilla arms cache. It arrives some hours later, carted by eight load-bearers, and Parsons, covered with leaves woven into his netting and camouflage grease smeared over his face and hands, silently leads his force down the hill towards the bridge. Springing out of the forest, he surprises the four posted sentries of the 82nd Airborne and 'kills' them. He orders his men to take up ambush positions to cover him as he and another Green Beret wire the demolitions to the ninety-foot steel and concrete supported bridge crossing a wide gorge. After it has been 'blown up' an Army Corps of engineers credits the operation with achieving a 10 per cent destruction of the bridge, which would deny its use to the enemy for about a week.

A coded message is received from Army HQ ordering Parsons to exfiltrate the area to deliver a personal briefing to his commanding general. A rendezvous is arranged with a helicopter which pulls him out and over the trees in a 'stabilized extraction system'. Harnessed to a rope, Parsons hangs in the air ninety feet beneath the helicopter for the hour's ride back to base. Met by a jeep, he is then immediately driven to the briefing room, where he delivers a forty-five-minute verbal situation report to the assembled officers, pointing out the areas where the 'enemy' is most vulnerable. The next day Parsons is sent parachuting back into the forest with an airborne battalion to act as pathfinder and link up with the friendly guerrilla forces as the invasion begins.

When Bo Gritz underwent special forces training, he was ordered to infiltrate the small town of Chester, North Carolina under a false identity. His undercover mission was to develop a network of clandestine agents and establish a communist-type system of subversive cells. He obtained a local driver's licence under the name of Jimmy Wilkes, worked in various odd jobs, gardening or flipping hamburgers at the local diner, before getting the job of deputy sheriff. He was in a strong position with several agents already recruited by the time an Army counter-intelligence team came into town to uncover him.

Gritz had told the owners of the local motel to be on the lookout for 'men in three-piece suits carrying briefcases'. When he was informed that suspicious strangers fitting that description had just checked in, Gritz immediately went over to see the motel manager. Having convinced the manager that his new guests were from the Mafia, Gritz obtained the motel's cooperation in his 'sheriff's investigation'. Allowed to search a room shared by two of the men, Gritz confirmed that they were part of the counter-intelligence team he was expecting, after finding Army documents among their papers. He bugged their telephones and had them all closely watched before they even knew that he was the deputy sheriff.

Anticipating their every move, Gritz deliberately drew their attention to the sheriff's secretary, who was one of his agents, by staging suspicious comings and goings from her home. Having arranged for several people to turn up at the house one night for what appeared to be some sort of clandestine meeting, Gritz made sure that they all got out through the back garden as the counter-intelligence team, still in their three-piece suits, started moving around the house to try to see and hear what was going on inside. As they attempted a surreptitious entry through one of the doors, a call immediately went out to the sheriff's office from one of the neighbours and as they heard the sirens of Gritz's patrol roaring towards them, the counter-intelligence team quickly jumped in their car and fled the town at full speed.

'Special Forces training teaches you individuality,' explains Gritz. 'You learn how to operate under your own authority. It becomes a matter of: don't tell us how you do it, just do it.' British SAS commander Michael Rose agrees that 'a special operator needs to be able to take an unpredictable approach and work his way around a problem instead of confronting it head on. If the enemy does not know what you are all about, they make themselves vulnerable to you.'

A British Army unit known as the 14th Intelligence Company is trained for specialized service in Northern Ireland. Its recruits come from all different sectors of the armed forces. Many have failed the physical tests of SAS selection but are found to be psychologically suited to special operations work. They are also required to have experience serving in Northern Ireland and useful work skills, for a trained Army clerk, cook or mechanic can find a normal job more easily and blend better into a civilian community than, say, a paratrooper. Women are also recruited into '14 Int'. Before conducting deep covert penetration of IRA-controlled communities, 14 Int operatives receive extensive psychological conditioning and in-depth schooling in Irish history and language, including Celtic dialects and accents. By the time he or she has finished the course a 14 Int operative should be capable of recognizing a man's regional origin, and even his political and religious orientation, by the way he speaks. He will also be trained in the habits and normal behaviour of a working-class Irishman. He will know, for example, when drinking at a pub, always to leave his empty beer glasses stacked on the table instead of taking them back to the bar to be refilled as do the English.

Another part of the psychological conditioning consists of memorizing complex verbal instructions and keeping up a detailed observation of close surroundings at all times. In one training exercise, a plain-clothes operative walks up to a call-box. It's exactly five minutes past the hour on his synchronized watch as he opens the door and the telephone inside rings. He is given the instructions about where to make his next contact and the caller abruptly hangs up. They are relayed only once. He walks several blocks to the location assigned for the rendezvous and is then questioned. 'There was a car parked right outside the call-box. Give us the make, registration number and other identifying features.' He is expected to remember, and if he lies in any way he is failed.

After they have undergone specialized training at 14 Int's secret base in Kent, recruits are put through a rigorous course of resistance to interrogation and learn the most lethal forms of hand-to-hand combat and close quarter battle (CQB). They will rely almost entirely on their bodies and a 9mm automatic pistol for their protection or to carry out the occasional hit job which the unit conducts against IRA gunmen.

A nerve centre located somewhere in Northern Ireland coordinates 14 Int operations, receiving coded reports from the various operatives working clandestinely throughout the six provinces and ordering an ambush when the proper intelligence is received. Some 'shoot-to-kill' operations conducted against IRA terrorists attributed to the SAS have actually been 14 Int jobs. In 1981, 14 Int became involved in an incident near the Northern Ireland Border in South Armagh. Driving a plain-looking but armour-plated sedan the 14 Int diver cruised along the crossing point that morning. He had taken the added measure of fitting the car with Republic of Ireland number-plates to make it an even less suspicious and more attractive target for the terrorists – an Irish-registered car would be harder to trace in Northern Ireland.

When the 14 Int driver saw two men flagging him down, he came to a slow stop, gripping the 9mm pistol loaded with a thirteen-round clip concealed on his lap under a folded newspaper. As the IRA men ran up to the door and produced weapons, he shot through the window, instantly killing the one closest to him with two bullets through the head. Crouching behind the armoured door as the other gunman, nearer the front of the car, returned fire, he pushed it open, knelt on the ground behind the cover of the door and, aiming through the smashed window, pumped out the remaining eleven rounds in his magazine, grouping most of them in his assailant's head.

The mastering of deceptive techniques is of key importance in training special forces, who are likely to find themselves facing a vastly superior enemy in a 'denied area', operating in a twilight world where military, police, intelligence and subversive roles blend together. 'Special Operations are special,' says Bo Gritz, 'because there are no textbook guidelines or classic cases to draw from in meeting the challenge of doing something which hasn't been done before. They require original thought from a mind that is not restricted or encased by normal perceptions of what can or cannot be done. Each new invitation erases past concepts and becomes its own universe to be analysed, organized and dealt with.'

When he led Green Berets in a guerrilla exercise against regular troops in the jungles of Panama, Gritz had flyers printed and surreptitiously distributed among 'the enemy', warning that the area was infested with deadly bushmaster snakes, a form of large, venomous viper. Everyone was advised to build big fires at night, not to sleep on the ground, to use only established roads and trails, and scuff their feet as they walked, to scare the snakes away. The notice was signed by the Commander in Chief Southern Command, the top US military authority in the American-controlled Canal Zone, which quickly disavowed it, but not before it had been read and taken seriously by many of the troops. As a result, their movements and locations were easily detectable by the Green Berets, who were constantly able to evade them at great distances and to prepare ambushes with plenty of advance warning.

When the turn came for the Green Berets to act as the enemy for SEALs, the dirty tricks got nastier. 'We played all kinds of head games on them,' recalls Richard Marcinko. The Navy operators would sneak up on their Army counterparts at night, tying them up in their hammocks and stealing their weapons, food and even their wallets and clothes. They wrote nasty letters to wives and girlfriends whose pictures and addresses they found.

Continuation training only gets worse. Most special forces units require their men to undergo a survival phase in which they are left practically naked and without tools in the middle of some wilderness to live off the fat of the land. The SAS might equip a group of about five trainees with nothing more than a knife and one box of matches. In some units, the men don't even get that.

The Bandera de Operaciones Especiales de la Legión (BOEL), which has grown out of Spain's Foreign Legion, conducts an especially tough survival course in autumn weather along the Sierra Nevada mountains above southern Spain's Costa del Sol. The men must live almost like their prehistoric ancestors with nothing but sackcloth and coarse blankets with which to cover themselves, and manage with no tools of any kind. They have to carve their own instruments for hunting and fishing from sticks and stones, and fashion a leaf-covered refuge with whatever they can find in the rugged hills. To build a fire, the Legionario has been instructed to drill a small hole through a thick slice of wood, fill it with dry grass and rub a stick through the hole until the friction creates sparks that ignite the grass. He can now boil soups from the various herbs and edible plants he knows how to identify amid the local vegetation – mint, nettles, dandelion, pennyroyal, cress, camomile. He can also catch frogs and cook them, but he will have been trained to tell the difference between different types, for some have poisonous skins. His normal menu also includes lizards, snakes, turtles, snails, worms and even certain protein-carrying insects. In addition he can catch fish from a stream or scavenge for a dead bird.

The survival course lasts about two weeks, or as long as it would take a man to starve to death. Instructors regularly drop by to check on the men, who are advised to keep physically active, despite their weakened condition, so as to avoid complete mental paralysis. Sugar, in particular, is missing from their diets, causing some to mentally hallucinate about sweets.

The Australian SAS leaves its men on some deserted offshore island or remote desert in the outback. Once they have been sufficiently starved, animals are brought in for them to slaughter. 'We will pick the youngest and most sensitive in the group and hand him a knife to kill a baby lamb by slicing its throat,' says an SAS instructor. 'It's also another way to test whether a man can cope with the feeling of warm blood pouring on to his hands.' If they are in a wilderness area where there is a threat of crocodiles or dingoes, the survival group is issued with a single rifle and a small supply of ammunition. In one famous episode a crocodile crawled on to a rubber boat which two Australian SAS men on a survival course were towing along a shallow river. They just kept pulling it along until they got to the shore and the man-eating lizard crawled back out.

In some cases the men are left to make their way back home on their own. One SAS sergeant recalls having to walk and hitchhike for over 200 miles covered in nothing but sackcloth, looking like some marooned derelict or escapee from a lunatic asylum. When he finally turned up at his doorstep, his wife hardly recognized him. 'When she finally opened the door, she shrieked with horror at my totally emaciated state. "What have they done to you now!" she cried. I hardly had a penis left any more.'

But most special operatives find that the worst experience of their training is the mock 'capture and interrogation'. One walks into the bonfire-lit courtyard of an abandoned farmhouse to witness the

Dantesque spectacle of some thirty blindfolded BOEL Legionarios, spread-eagled against the surrounding stone walls, their cries and moans bespeaking the worst physical and mental tortures possible within the confines of legality. Instructors are forcing some to do one-handed push-ups and remain resting sideways, suspended on one fist. Others are being jostled, kicked and shoved. Every fifteen minutes or so they get pulled around the courtyard in a single file, each man holding on to the next as if partaking in some satanic dance. Around and around they go until some drop off from dizziness to be violently pulled up by one of the sergeants and pushed back up against the wall.

One by one they are fed into the interrogation room, where a man is held down on a chair while clear plastic is wrapped around his head. As he starts suffocating, a hole is plugged into his mouth and he is told, 'Now talk.' He is not supposed to reveal anything except name, rank and serial number, according to the rules of the Geneva Convention. A long stick is banged on the table in front and on the back and sides of his chair as the instructors warn him that the next whack will be across his face. The NCOs scream out insults and air personal grudges to make the men believe that there are real motives behind their brutal treatment. One blindfolded Legionario with whom I am briefly allowed to speak says that the worst part is not being able to see. Before his 'capture' the BOEL operative had been through an overnight twenty-kilometre mountain march and a simulated combat exercise, crawling beneath barbed wire under live machine-gun fire and exploding demolitions charges.

In an SAS training interrogation, the men are stripped completely naked and forced to overhear the terrifying theatre of mock tortures going on in an adjoining room. The sound of a man screaming in pain and crying for mercy as a mattress is beaten can be dreadfully realistic when one is blindfolded, frightened, fatigued and in a confused state of mind. To humiliate them further, the SAS forces its trainees to sit tied to a chair, totally naked, as they are interrogated by a woman who makes degrading comments about their manhood. This would undoubtedly be unacceptable to a Spanish Legionario's machismo.

'You feel terribly vulnerable when you are blindfolded,' writes Sergeant Andy McNab of Bravo Two Zero, the SAS patrol which was compromised inside Iraq during the Gulf War, with the result that four men were subsequently captured. He remembers how being able to peek through his blindfold at his surroundings during his brutal interrogation by Iraqi Intelligence agents gave him a tremendous sense of reassurance. His SAS conditioning paid off. It took eight days of constant beatings, having his teeth ripped out and cigarettes stubbed on his skin to extract any information out of Andy, who still didn't tell them the whole story. He revealed that he was part of a British Army reconnaissance patrol observing military movements inside Iraq but never mentioned the SAS or their real mission of blowing up Iraqi underground communications lines and Scud missile emplacements. 'I had to hold out long enough to make them really work hard for my cover story, to make them really believe that they had achieved something by dragging it out of me.'

During the Vietnam War, US Army Special Forces built a mock Viet Cong prison camp near Fort Bragg where Green Berets would be imprisoned for up to a week under fully realistic conditions. In this fantasy setting the men are locked up in open cages and fed nothing but balls of soggy rice while being harangued with propaganda broadcast over loudspeaker systems. The prison-camp model is based on the written experiences of Captain Nick Rowe, the first Green Beret to be captured by the VC, in 1963. Despite the worst imaginable physical tortures and deprivations, they never broke him mentally and he managed to escape after years of captivity.

The Australian SAS runs a similar facility in an old fortress near Sydney, where special forces trainees are deprived of sleep while being kept in partially flooded, rat-infested cellars. During interrogation they are forced to kneel naked with their knees resting on a broom-stick and both arms stretched out. They are disqualified if they reveal as much as their place of birth. The date is already on their dog-tags.

The prototypical Hollywood image of a special operative as a fast-firing Rambo devastating all opposition before him in an explosive rush of death and destruction may have its time and place. But the usually silent job of reconnaissance and surveillance behind enemy lines, which is the most common special forces role, requires careful learning so as to avoid contact with the enemy. British Royal Marines of the usually low-key Special Boat Service (SBS) take considerable pride in having avoided contact with Argentine forces throughout the Falklands War. One SBS lieutenant tells me that he feels they did a more professional job than the SAS, whose OPs were compromised on a couple of occasions, leading to the death of one SAS officer near Port Howard on West Falkland and the capture of his radio man. An SBS OP, on the other hand, positioned on a main supply route where it was being continuously overflown by Argentine helicopters, was never detected.

During long periods of realistic training, the SBS, SEALs and other units are conditioned though 'fire control' exercises in which they are marked down if they give away a position by shooting back on either hearing or seeing gunfire. They are taught to be able to tell when shooting is directed at them and when it is 'reconnaissance by fire' in which enemy troops shoot randomly to see if there is anybody out there. Special operatives learn that they can often crawl right through an enemy position in the dark if they maintain strict silence and discipline.

Four SEALs, dug into a sand dune on the border between Saudi Arabia and Kuwait, were reporting on an Iraqi armoured column moving south on the road towards Khafji during the start of Operation Desert Storm. They were the first to spot the only major Iraqi penetration of Allied lines during the Gulf War, getting word out to headquarters a full four hours before the Saudi town was actually attacked. Suddenly the lieutenant commanding the patrol realized that three armoured personnel carriers had separated from the main column and were moving across the desert towards their position. 'We had probably been picked up through a radio intercept.'

Heavy machine-gun fire stitched its way towards the dune. Although equipped with rockets and grenade-launching 203s, the SEALs held their fire as it would give away their precise location. Taking advantage of the few minutes they had before the tanks closed in on them, they just packed up their radios and the rest of their equipment and had quietly made their way over the dune by the time 20mm cannon fire started ripping into what had been their hide. They escaped into the desert, linked up with a unit of Marines and turned up at their headquarters at Ratt al Mishab after being out of radio contact for almost six hours. 'They were cool as cucumbers,' recalls their commander, Captain Tim Holden, 'as if they had just come back from having dinner for a debrief.'

A team of the US Marines' 1st Force Recon Company, which trains in a similar way as the SEALs, was positioned on the road to Khafji as the Iraqi tanks rolled in. Radio reports which had been received about the SEALs' border position taking direct fire had diverted them from their original target that night, an Iraqi frog missile site across the border. The Marines made their way into Khafji instead, allowing themselves to be surrounded by the Iraqis as they took up an observation position on a rooftop from where they guided in air strikes during the entire twenty-four-hour occupation of the town. They were never detected while reporting closely on the situation in Khafji. When sudden bursts of gunfire could be heard coming from within the town the Marine OP confirmed that Iraqi soldiers trying to defect were being shot. When Saudi troops retook the town the Marines continued guiding them from their OP in house-to-house searches.

Training in OP work mainly consists of learning to operate and survive in underground hides, where life is about as pleasant as the worst enemy prison camp, even with the latest conveniences. The only thing that should be protruding from the hide once the men are dug in is a field periscope moving up and down through the twenty inches or so of earth covering its roof. 'Spending four days underground with three of your oppos can be incredibly boring,' reports one SAS man. 'The main thing that actually came out of the exercise was how appalling this role really would be because we were stuck down in this underground shelter for three solid days and nights in the most squalid conditions imaginable.'

Recently some comfort has been lent to the ordeal by a prefabricated shelter made of corrugated plastic with a steel frame which lines the hole like an inverted tent, insulating the men from the earth, dampness and rats. But this does not alleviate another serious problem. 'When Royal Navy helicopters came to pick us up and we moved into our helo, the RN crew physically drew away from us and kept us at arm's length. We couldn't figure out why at first. We had been out in the field for five days and in the underground shelter for four. We had no idea how disgusting we smelt because we were used to it. We smelt incredibly awful. In the shelter, we had to crap and piss into different plastic bags which were then sealed together in one big plastic bag. Do both together in the same bag and it is prone to explode. Before this exercise, they tended to explode inside the shelter, a problem which has since been alleviated, but you could end up covered in the stuff and with no real way to clean up the mess. We were also "hot-bunking", using two sleeping bags between the four of us, surrounded by all our kit with no room to move. The food was appalling and, probably because of the stench, tasted awful. I'll never forget when we at last came out. The first gasp of fresh air after ninety hours underground – it was quite unbelievable.'

Another SAS man experienced in OP work whether in the field or in an eighteen-inch bricked-in corridor between terraced houses also describes how previously unnoticeable traits and minor habits of his 'oppos' become magnified in the cramped conditions. 'You suddenly become aware that one of your mates is a nose-picker. The dust gets absorbed into the mucus so that you hear his teeth gritting into the mix as he puts it in his mouth and it really starts to get on your nerves.'

The US Marines' 1st Force Recon Company solves the problem with a system of interlocking individual hides. Each member of a four-man patrol digs his own hole, laying a sheet inside and scraping a small tunnel through which he can communicate with the next hide, a foot or two away. This tackles the problem of stench, as the excrement of all four men does not get concentrated in one hole, and it allows some privacy. The Marines also maintain that smaller holes are easier to conceal.

'Hello, Zero. This is Alpha Three. Contact. Wait out.' It's the signal for all other SAS patrols to get off the radio net for Alpha Three to call in fire on an enemy target from their OP. 'Fire Mission. 14.00. Grid reference 444. Three T-72s moving south at eighty degrees approximately one kilometre from our position. Fire for effect.'

A mortar battery about ten kilometres away immediately lets off a barrage of 81mm shells which fly 10,000 feet into the air before crashing down about 100 metres from the tanks. Watching it through his periscope, the commander of Alpha Three Zero gets back on the radio to redirect the fire slightly east. Heavy artillery positioned thirty kilometres back corrects bearing accordingly and a hail of 105mm rounds come screaming down less than thirty seconds later.

'Target on,' reports Alpha Three Zero as one tank explodes from a direct hit. 'Bearings now seventy-five degrees south-south-east,' the OP reports the changing direction of the surviving tanks which are trying to take evasive action.

'Ten 105s coming your way,' the artillery position informs the OP of the quantity and type of rounds being fired which fall some seconds later with full accuracy, destroying one tank and severely damaging the other.

'Hello Zero. Alpha Three Zero contact report. 14.05. Grid reference 444. Three T-72s launched attack. All destroyed.' Further details, such as estimated enemy casualties, will be encrypted and sent by burst transmission on the Plessey PRC 319, the most advanced field radio now in common use with the SAS and other special forces. A key pad for entering data which can be detached from the main set is being punched by two men at a secondary OP some distance away as they see more enemy tanks lining up. Their four-line message giving the number of tanks, their bearing and other information is sent in a short burst lasting one-tenth of a second whereas it would take several seconds to transmit by voice. Another software ingredient called the Special Operations Radio Frequency Management System, incorporating interference receivers which can instantly predict the best radio frequencies, is automatically switching the ongoing transmission on to its selected bands.

Until the new American-developed Joint Advanced Special Operations Radio System (JASORS), SAS radio operators had to learn to manually change frequencies every few seconds during transmissions to avoid enemy interference. The technology now available to the men shitting into bursting plastic bags inside a hole no longer makes this necessary but every member of the SAS must become acquainted with the use of about thirty different types of field radios and be able to disassemble and reassemble them. When it comes to calling in artillery or air strikes against a vastly superior force, operating a field radio becomes about as important as handling a weapon. Indeed it was largely because of the failure of their radio systems that the Bravo Two Zero patrol became lost in Iraq.

In addition to their new burst-transmission radios, an elite special forces patrol operating a sensitive mission deep into hostile territory might also go equipped with a TACSAT, a lightweight, portable, long-distance, ultra-high-frequency satellite radio. Also equipped with an embedded encryption system, a man in the Vietnamese jungle can use it to communicate directly with his central command in the Pentagon.

The bewildering new array of special operations technology now also includes the Improved Remotely Monitored Battlefield Sensor System (IREMBASS). This is a box with magnetic, seismic, acoustic and infrared sensors and relays which can be concealed in a 'hot' location by a patrol and monitored from a safe distance of up to seven kilometres. It can detect precise details of enemy movements, including types and numbers of vehicles moving along a road or aircraft landing and taking off from an airfield.

Thermal night vision goggles (NVGs) now give far better depth perception and a clearer picture in the darkness than the previous infrareds with their greenish world of hazy heat-emitting objects. Yet a newer generation of vision-enhancement equipment is being developed which can detect human heat sources behind most objects, including trees, foliage, earth and most walls. Only glass, aluminium, some plastics and water can deflect the new heat-seeking rays. There are laser target designators which look like oversized binoculars or big, squat telescopic rifles and fire invisible rays which 'paint' the target. What appear as red dots on the LTD's cross-hairs can be simultaneously picked upon the hood screen of an aircraft flying thousands of feet above, directing air strikes on ground targets with pinpoint accuracy. Similar laser devices are now in use to direct mortar fire.

It is with the near certainty of being able to see and hear almost anything or anybody, at any distance, day or night, that today's special operator launches into dangerous live-fire practices which few other military units dare to engage in and many senior officers find downright frightening. 'Hey! wait a minute! There are real people down there. I can see a vehicle, aircraft and all sorts of equipment. Is somebody else using the range?' asks the general who is about to launch a special forces assault on a mock missile installation built during the sixties when Green Berets were practising taking out the Soviet long-range missile emplacements in Cuba.

Bo Gritz has added deceptively real dummies and man-sized silhouettes with faces and uniforms painted on, around a missile on its launcher with its support van, a radar, tents, jeeps and an old grounded Choctaw helicopter. He just begins explaining this to the general when 60mm mortar shells start hitting the target. The thud of falling shells and black puffs of smoke are clearly seen and heard as the twelve-man assault team gets into position. 'My God! What's going on now? The mortars are going to hit those men. Stop this!' The general takes his eyes off his binoculars and turns to look at Gritz. 'It's against Army training policy to use indirect fire in close proximity to troops.'

The A team penetrates the installation, blasting holes through the perimeter's barbed wire with Bangalore torpedoes and firing grenades from 203 launchers attached to their M-16s. As in a real combat operation, the mortar bombardment had ceased just as the assault team got close enough to put down their own suppressive fire. But since there is no noticeable break or much difference between the mortars and the grenade explosions, the general thinks that the bombardment is continuing as the men run inside. 'This is illegal,' he says, visibly shaking.

After hosing down all the personnel targets with their automatic rifles, the A team places 40lb of explosives around the missile site, wiring the demolitions charges and clearing the area in just a few minutes before the ground shakes with one big tremendous explosion. Soil and vegetation shower over Gritz and the general as the missile, its launcher, the helicopter, radar, van and other objects are vaporized. When the general gets back in his jeep, he is too shaken up to reprimand Gritz on the explosive overload. There is a maximum limit of 4lb of TNT on the practice range.

Live-fire exercises are common to other elite units. Paratroopers, Rangers, Marines are all front-line light infantry which may be called on to directly support a special forces direct-action mission. While a team of SAS, SEALs or Delta will be sent to deal with a central target inside an enemy territory – a missile installation, headquarters, hostages – Rangers or Marines might be inserted to take peripheral targets or 'create an environment in which special forces can work'. They might cut off routes of enemy reinforcement to the central target or hold an airstrip or beachhead for the exfiltration. 'A man's natural instinct under live fire is to duck for cover and do nothing,' observes one American Ranger officer who has also served with Delta. 'It's only by experiencing a situation where hot lead is actually flying all around that you condition soldiers to return fire and manoeuvre in an intense combat environment. You can't achieve this with blanks.'

The more specialized the unit, the more complex and dangerous live-fire manoeuvres become. Rangers will move on a target from its flanks while machine-gun fire is directed on to it from the front. SEALs will practise withdrawing under live fire with one section of a patrol shooting on to an oncoming enemy from covered positions while men further in front pull back through the hail of bullets and grenades.

The SAS practises an ambush with live ammunition, claymore mines and anti-tank weapons. In a linear ambush, a troop subdivides into four sections composed of four men each. The 'killing group' in the centre is flanked by 'cut-off groups' on either side and a 'rear protection group'. The groups are out of each other's view when a flare goes up as the signal to detonate the mines along a road or footpath and they all open fire. In the course of the ambush, two men may rush towards a vehicle underneath the whoosh of a Milan rocket flying over their heads which explodes on hitting the truck in front of them.

In a counter-ambush, eight men move quietly several yards apart in a wedge-shaped formation through thick bush country. It's night and they are seeing through NVGs, which limit their peripheral vision. Suddenly a pneumatic target pops up in front which the lead gunner instantly drills with bullets from his automatic rifle. The flashes and sounds of gunfire go all the way down the line as more dummies appear around the sides and behind them and each man on the patrol covers his arc of fire.

Looking out from the church steeple in the picturesque town of Mont Louis, set against the snowcapped Pyrenees, Corporal Peter Cuxson, his face smeared in camouflage grease, peers though the sights of his FAMAS rifle. He can see the woman who comes out on to the balcony across the street with a pot of fresh coffee to enjoy breakfast in the crisp morning light. But no sooner does she catch sight of the sniper from the CRAP (Commandos de Renseignement et d'Action dans la Profondeur) platoon of the 2 REP (Deuxième Regiment Etranger de Parachutistes) than her fine china cup drops to the floor and the steaming black liquid spills around her feet as she is seized with terror.

Neither the mayor's wife nor anyone else in the village was aware as they woke that spring day, that their town had been taken over during the night by the 2 REP. All entry points into the village were sealed off with checkpoints manned by camouflaged paratroopers with their distinctly shaven heads. Armed jeeps stood in the village square. Cuxson as well as other snipers and machine-gunners commanded the heights of the village.

With their coup accomplished, the Legionnaires regroup and proceed to march through the old cobblestone streets singing the French Foreign Legion's hymn with lyrics borrowed from the SS: 'We ask the devil to march with us, we are soldiers for pay.' It was a nice way to end an exercise. 'We got away with things in realistic training that no other unit would get away with,' says Cuxson, a British ex-Para who joined the Legion after serving in the Falklands War. Accepted by the 2 REP despite a twisted spine, he trained in demolitions, sniping and infiltration by air, land and water to become part of the Commando de Nuit. Later he was selected to become one of ninety specially trained Legionnaires in the CRAP, the very pinnacle of the Legion.

Considered one of the most professional French special operations units, the CRAP's selection is as physically and psychologically strenuous as that of any other special force. It requires a thirty-kilometre forced solo march through the Pyrenees loaded down with a 27kg bergen. Struggling through an obstacle course in a ruined castle – where the SAS is also invited to practise – 'CRAPS' must jump over a 1.5-metre trench between two battlement walls. Several have been killed trying it. Capture consists of being ambushed and beaten while on the way to a routine shooting exercise and thrown into a dungeon from where a man has to escape blindfolded through a hidden trapdoor before being captured, beaten and interrogated again.

Demolitions training includes rolling on to a road in front of a moving tank, getting between the grinding tracks as it rolls over and attaching magnetic limpet mines to the bottom before coming out the other end. Cuxson also learned to steal tanks and other vehicles by pumping air out of a small hole in a tennis ball to pressure open the central locking system. In one realistic exercise he was required to steal a motorcycle. Sabotage training gets highly technical at the 2 REP. Demolitions experts are taught, like engineers, to calculate the stress levels of different buildings and structures in order to apply the precise amount of explosives to produce the desired level of damage.

The 2 REP also practises hostage rescue by abseiling on to a rooftop from a flying helicopter and blowing frame charges through the upper windows and hatches. But in a competitive counter-terrorist exercise with the even more specialized SAS, Cuxson admits that 'they beat the crap out of us – no pun intended'. During a jungle training exercise with the SEALs in the former British South American colony of Guyana, however, it was the 2 REP which beat the US Navy's special operatives. 'They had great equipment and were excellent at operating in the water but inside the jungle they tended to get lost. They also relied too much on their M203 grenades which lose effect in the thick forest because they have to travel for twenty metres before arming themselves and would often hit trees, falling like duds,' says Cuxson.

'The training at the 2 REP is as intense and diversified as in any other special forces unit which I've come into contact with. They learn how to handle and operate all different types of foreign weapons, which comes naturally to a unit composed of men of eighty-one different nationalities, most of whom have previously served in their respective armies. The 2 REP has been drawing a great many recruits from elite units in Eastern Europe since the dissolution of the Warsaw Pact, including some Russian Spetsnaz, gaining more expertise in Soviet weaponry and techniques than any other unit in NATO.'

Like the SAS, in which squadrons are divided into differently specialized troops – Mountain Troop, Water Operations Troop, Special Vehicle Troop and Airborne Troop – the 2 REP is divided into four specialized companies: Commando de Nuit, or Night-time Commandos; Commando de Montagne, the Mountain Commando, trained in mountaineering, combat skiing and Arctic warfare; Water Operations company, composed of swimmers trained in underwater diving; Sniper-destroyer company, 'who spend all day shooting at targets and blowing things up'. The Sniper-destroyer company are all carefully documented by the French government since any of the highly trained killers could easily go on to become freelance hit men.

As does any other elite special forces unit, the CRAP cross-trains its men in all of the different specializations. Cuxson has qualified in everything from high-altitude free-fall parachute jumping to underwater swimming with scuba tanks and rebreathers. After learning to remain for prolonged periods at depths of no more than thirty metres, below which the rebreather's oxygen-recycling device stops working, he is sent on an eight-kilometre swim, navigating by compass to an assigned infiltration point. Emerging about 100 metres from the shore, he swims through the rough surf with his automatic rifle slung over his back and towing about 14kg of equipment. Reaching the beach, he has to climb a 200-metre cliff with a grappling-hook, to an RV point where he links up with a CRAP team which has been inserted by parachute for the toughest phase of all, the 'coup de main'.

They must penetrate a highly guarded 'enemy' installation protected by high-tech booby-traps and electrified fences. Infiltrating into a sensitive area today is infinitely more complex than at the time of David Stirling's lightning raids on German airfields. Approaching their targets, the special forces team might get detected by IREM-BASS, which would give the enemy plenty of time to set up an ambush aided with night-vision equipment from which the raiders would have little chance of escape. They move with sufficient stealth and speed to outmanoeuvre the sensors but then have to pass through a field of hidden plastic-encased land-mines. Using a hand-held 'sniffer', an ultra-sensitive scanner which can detect a spoonful of sugar at the bottom of a swimming pool, the commandos carve a path through the minefield.

Throwing their rubberized plastic ponchos over a section of the electrified perimeter fence, the CRAPS climb over the three-metre-high obstacle, going over one another's shoulders until the last man is pulled up with the demolitions equipment. But then they stumble across an invisible microwave fence with infrared cameras which can flash an immediate signal to the guardhouse, setting off an automatic alarm. Detecting the infrared beams with their NVGs, the men work their way around the high-tech barriers.

Meanwhile a CRAP sniper-destroyer positioned on high ground a couple of kilometres from the installation is taking careful aim with a .50-calibre rifle at the installation's emergency electrical generator. With the red centre dot of his infrared telescopic sight firmly set on the junction box of the power outlay system, he pulls the trigger of his bipod-mounted gun fitted with a sound suppressor. The half-inch armour-piercing bullet makes a puff of smoke visible through the telescope as it hits the target. The sniper knows that his part of the mission is accomplished but any sentry standing near the now incapacitated generator would have barely heard the metallic zap.

A few kilometres away a synchronized raid is being conducted against the main electrical power station servicing the base and its surrounding area. After lying up all day in a concealed position, the six men have redone their camouflage grease, applying fresh smears of green and black over their faces, necks and hands. Electrical tape is wrapped around their boots to muffle the sound of the buckles. After checking and rechecking their explosives and lightweight 5.56mm assault rifles, they bury their backpacks with all non-essential equipment and move on to their objective, dividing up into three two-man teams to execute different phases of the operation. Team 1, led by the platoon commander, will enter the power station to plant explosive charges on both transformers. Team 2, commanded by a sergeant, provides sniper support and sentry elimination. Team 3 will cut communication lines along the access road and seal off the target.

All three groups crawl for the last kilometre until they are within a few metres of the perimeter fence. Having studied close-up satellite photographs of the power station, they all know their assigned points. The sniper team spots a sentry patrolling outside the fence. Aiming his rifle, the sergeant covers the wiry Corporal Carbonell, who crawls up behind the guard and pounces on him, tightly wrapping a choke wire around his neck until he hears the crack of the broken respiratory tube, and then drags the dead body into the undergrowth.

The sniper team leader immediately reports the sentry's elimination to the squad commander, whispering into the ANPRC-77 radio transmitter clipped to his shoulder and wired under his shirt. When the lieutenant hears seconds later from Team 3 that they have cut the telephone lines and are watching the approaches to the power station, he immediately crawls with his other man up to the perimeter fence and cuts a hole in the wire with a pair of pliers. At the same time, Corporal Carbonel climbs a tree, getting into an elevated position from which to cover the demolitions team reaching the two main transformers.

Spain's BOEL, which exercises frequently with the 2 REP, prefers to combine plastic explosive with TNT in their demolitions charges. Legionarios of BOEL, which operated over two weeks without being detected in a German military zone during a competitive NATO exercise in 1989, explain that plastic tends to lose potency when exposed to damp weather over prolonged periods. TNT, on the other hand, remains stable. The lieutenant and the other demolitions man each paste down a 300-gram block of TNT using an equivalent quantity of Goma Dos plastic on the steel surface of each transformer. They inject aluminium starters into the charge, connecting its cable with the detonator fuse, and string out the electrical wire as they exfiltrate the plant through the hole in the fence. From the top of his tree, Carbonel catches an approaching sentry through his infrared sights, instantly releasing a muffled shot which hits the man in the forehead.

Reaching a distance of just over ten metres from the fence, the demolitions team connect their fuse wire to a crank-operated detonator which melts as it generates the electrical current to set off the charges. The explosions, raining sparks and yellow-pink flashes as both electrical transformers are totally destroyed, are the signal for all three teams to clear the area as fast as they can towards a prearranged RV point. Less than ten minutes have passed since the first sentry was taken out.

With searchlights and detection and alarm systems now blinded inside the main installation, Cuxson and his CRAP unit can move with relative ease to accomplish their mission and exfiltrate the area without being detected. They have several hours before the damage to the emergency power generator, which was disabled by their sniper, can be discovered and repaired.

Cuxson was once part of a 2 REP swimmer team which was preparing for a highly secretive special mission for the French government. 'We went into quarantine at the Legion's base in Bonifacio, Corsica, for intensive training in blowing up a boat. The mission required approaching the target at sea, and attaching explosives to the hull. After six weeks of constant practice we were stood down with no further explanation.' Three weeks later Greenpeace's ship *Rainbow Warrior*, which was trying to block French nuclear weapons testing in the South Pacific, went down off the coast of New Zealand. The French military intelligence organization, DGSE, had chosen at the last minute to entrust the operation to the Commando Huber, the special unit in French Navy's elite Nageurs de Combat.

Philippe Legorjus had been a stone-throwing radical during the 1968 student uprising against President De Gaulle. He was among the planners of the Paris street battles against helmeted riot police which were fought out with Molotov cocktails in the narrow, barricaded streets of the Latin Quarter. Memories of those days inextricably linked to his early experiences of free love and his beloved Chantal, the pretty blonde with whom he shared a small flat along with several other students, came back to him as he tested the impenetrable defences around the nuclear submarine base of Brest.

Dressed in jeans, sweatshirt and leather blouson, looking much as he would have when he was a student radical except for his close-cropped hair, Legorjus could have been a member of a dangerous terrorist organization. But he was now an officer in the Commando Huber. He had already been on endless routine exercises, jumping into the water from a helicopter with three other men and using a motor-powered Zodiac boat to make a night landing on some beach. 'Killing' patrols of the Gendarmerie, they would make their way inland to plant dummy explosives on some objective and then evade capture for the next few days. Legorjus would often do this by hiding away in the luxury of a friend's nearby château.

But this mission was different. He had to penetrate one of the most securely guarded French military installations in total secrecy, get into a nuclear-missile-carrying submarine and 'destroy' it. He and the four others in his team had checked all possible entry points around the complex but could find no weak links through the concentric rings of electrified perimeter fence, permanently manned guard posts and microwave alarm systems. Any commando approaching by land or sea would be discovered before getting even close to the submarines. Their only hope was to find some way of passing unnoticed through the main gate or docking area into the 'protected zone' which some civilian workers were allowed to enter. Legorjus had opened a file on all personnel employed in the submarine facility who lived off base while other members of his team observed the daily routine and security procedures of buses and ferry boats entering the base.

While Gendarmes systematically checked all special passes and IDs of civilians and military personnel going into the protected zone, the Nageurs de Combat did not fail to notice that the guards neglected to inspect the baggage compartment in the morning bus. They also tested people entering the base on civilian ferries, in which virtually no security precautions were taken. Legorjus looked up the morning bus driver's address and went to inspect the semi-detached suburban home. He found, much to his delight, another weak link in the security chain – the bus was kept parked on the street outside the house all night.

The commando team now drew up a plan. Legorjus and one other man would slip into the luggage compartment in the bus while the three others got on the morning ferry. They would then RV among the construction workers, congregating at 8 a.m. in front of the last checkpoint to be let into a building site near the berthing area for the submarines.

Sweating inside the cramped darkness of the luggage compartment, Legorjus counted each stop, having memorized how many the bus made before reaching the main gate. The bus stopped there a long time as the guards checked the IDs of each one of the thirty passengers. It seemed that they were taking longer than usual and Legorjus became afraid that the Gendarmes might have been alerted and could sniff them out with dogs. But the bus was finally let through. As it screeched to a halt after circling before the final checkpoint of the submarines' inner sanctum, the commando officer swiftly opened the boot from the inside and rolled out into the cold, grey morning with the other Nageur. The two men blended among the crowd of construction workers and spotted the three other team members, who had come off the ferry. Looking rough and unshaven, one of them had a lit Gitane hanging from his mouth as he pushed his hands deep into the side pockets of his anorak with its wide collar turned up against the damp chill.

They were now prepared to execute the critical phase of the plan. The team of Nageurs got into line just behind Legorjus and as his turn came for a pass check at the control post one of the Nageurs yelled at the one smoking the cigarette, '*Espèce de salud! Mange merde!* If I catch you with my wife again, I will break your face.' Then he pushed him violently. As the Gitane came unstuck from his lip, the man yelled back, 'Eh! If your prick's too short then I guess I'll have to fuck her myself, asshole!' Hands were now out of pockets and they came to blows. Another Nageur joined in the fight, causing a mêlée, and other workers got caught up in the scuffle. As all three Gendarmes at the checkpoint waded in to break up what was turning into a riotous mob, Legorjus slipped past the checkpoint, followed by another of his men, and they headed straight for the submarine docks. Seeing the crew of *Le Redoutable* standing to attention along the bow for the morning hoisting of the colours, the two commandos jumped on to the deck, entered the vessel through an open hatch, closing it above their heads, and climbed into the interior. They walked around the elevated bridge with its periscope tube, past the empty weapons-control and sonar stations, then Legorjus picked up the telephone. Dialling the number for Commando Huber's headquarters and being connected, he reported in. 'Lieutenant Legorjus here. The *Redoubtable* is in our hands. I await further instructions.'

Philippe Legorjus went on to become an officer in the French Gendarmerie Nationale, where in 1985 he succeeded Lieutenant Christian Prouteau as commander of its special counter-terrorist unit, the GIGN.

7

THROUGH THE KILLING HOUSE

Looking like an abandoned city amid the dry forests of Western Australia, stands the 'Killing Village' with its multistorey buildings, including an embassy, office block, semi-detached houses and shopping centre – all mock-ups. One of the world's most complete counter-terrorist training grounds, it was built by the Australian SAS and is used by special units from all over the world willing to pay generous fees.

At first nothing appears to be happening. A few vehicles are stationed in the car park with its yellow meters in front of the two-storey shopping complex. Broken glass and spent 9mm brass casings are strewn everywhere within the rubber-coated concrete walls of the various buildings. It has the feeling of an empty movie set. Then there is a mechanical noise, like a conveyor belt. Moving on rails leading out from the shopping centre are four robots. One is done up as a female complete with plastic breasts while the rest are males. Two are armed with what seem like real pistols pointing at the heads of the hostages. They stand for a moment near a red Toyota, then move back on the rails which disappear into the building.

Hidden up in twin towers which rise above the middle of the set, two pairs of SAS snipers have been observing this staged drama through their telescopic sights for days. They also see pneumatic targets popping up behind the windows of the shopping complex, and buzz central control over their walkie-talkies each time to identify if they are seeing a hostage or a terrorist. Lying on the hard grey steel floors, aiming German-made PSG-1 high-precision rifles which rest on small tripods, the marksmen are sure never to let the barrels protrude through the windows, their hands at all times firmly holding the grips, their forefingers brushing the triggers. Replacement rifle barrels are laid out to one side. The men in each pair alternate the watch, taking turns sleeping, eating and relieving themselves in plastic bags which are collected during the night.

They might end up lying there for the next week as the hostage scenario unfolds. By pulsing one of two buttons on his walkie-talkie, each sniper activates a green or red light in front of his assigned number on a Motorola-designed fire-coordination console which is closely monitored in the crisis control room in one of the buildings. Green indicates that the sniper has his target located and when it switches to red the control officer knows that the sniper has the target firmly in the cross-hairs of his sight and is prepared to shoot. As negotiations drag on with the terrorists and points of breakdown or danger to the hostages are reached, the control centre always knows the disposition of the snipers to open fire and the instant when they can synchronize a volley to take out all targets simultaneously.

'One shot and that's it,' is the ideal prescription for ending a hostage siege advocated by Philippe Legorjus of the GIGN. But only if all the snipers are on red and they have detailed and precise information on the exact number of terrorists and hostages and what each one looks like, would such an option be favoured. Otherwise, there are about forty men standing by in their fire-retardant suits, respirators and body armour, checking and recheck-ing their frame charges, 9mm pistols and MP-5s, preparing a 'distractive entry'. Once the direct assault is launched, the sniper team also goes into action, shooting at will on 'targets of opportu-nity'.

On the day of the May 1980 Iranian Embassy take-down in London, the Australian SAS was graduating its first fully trained counter-terrorist (CT) squadron. They watched the television news clips of the assault at an airfield in Victoria where they had just been put through their final counter-hijacking exercise by a five-strong team of instructors from 22 SAS. The very same men from B Squadron who were machine-gunning and flashbanging their way through the Iranian Embassy, had trained an original group of eight Australians in the black arts of close quarter battle at the first SAS 'killing house', built in Hereford back in 1976.

'What's it all about?' asked Sergeant Bob Mawkes when he was first put in charge of the Australian SAS CT programme.

'What do you think it's about?' his superior officer threw back.

'I don't know but it sounds interesting,' replied Mawkes, who went on to visit all of the fledgling counter-terrorist units starting up in the late seventies: 22 SAS in Britain, the GIGN in France, the GSG9 in West Germany, and Delta Force and Blue Light in the USA. 'I saw that there was a future in it,' says Mawkes, who takes pride in his Sephardic Jewish background and wrote a 180-page report proposing a training programme and facilities to create a fully-fledged CT squadron in the Australian SAS.

The multimillion-dollar construction contract for the 'killing village' was awarded to Australian Construction Services. The various buildings were equipped with interchangeable walls wired for moving targets and computer simulations of hostage scenarios for realistic indoor room clearings. The investment soon began paying off, literally, as 22 SAS, Delta, SEALs, Japanese police SWAT teams and other units paid good money to come to Perth and use what Mawkes describes as 'the most sophisticated CT training facility in the world'.

'Training had to take into account the kind of instruction being offered to international terrorists,' explains Mawkes. Intelligence shared between the various CT units in the early eighties indicated that Muammar Gaddafi had made Libya's national airline available for groups like the Palestinian extremist Abu Nidal to practise hijackings. 'We had to do the same in order to counter it,' says Mawkes. Qantas Airlines provided the Australian SAS with an airliner in which to train in anti-hijacking. British Airways did the same for 22 SAS, Eastern Airlines provided one of their jet liners for Delta, Lufthansa for the GSG9 and Air France made its entire air fleet readily available to the GIGN.

Counter-hijacking is considered one of the most dangerous and risky special forces operations. 'If you can master that you should be able to do anything,' says Philippe Legorjus of the GIGN, which has trained on every kind of airliner there is, from small DC9s to Jumbos. As a result of rehearsing counter-hijacking once a month, the GIGN claims to be the best at it. One exercise consists of abseiling by rope from a helicopter flying at sixty feet on to the fuselage of an airliner taxiing on a runway and gaining access to the interior through the escape hatches. Legorjus, Christian Prouteau and the GIGN's current commander, Captain Denis de Favier, all maintain that the French group has developed a secret method of penetrating a closed aircraft without using explosives. 'We have learned how it is possible to open a plane's emergency door from the outside,' elaborates Prouteau.

According to him, the GIGN has used this system on three occasions.

In one incident in which an IRA terrorist commandeered a plane at gunpoint from London to north-west France, the GIGN's main fear was of a booby-trap bomb which the hijacker had rigged inside the aircraft. Taking advantage of demands for food to be brought to the plane, four GIGN men gained access to the plane by pretending to be stewards carrying the pile of trays. Overpowering the gunman by quickly getting him in an arm and head lock and with a .38 revolver brushing his temple, they managed to disconnect the detonation device for the bomb, which he was carrying on him. When Delta once rehearsed such a deceptive method for entering an aircraft, the men playing the terrorists insisted that those bringing the food take off all their clothes.

While the standard counter-hijacking technique involves 'engaging down the fuselage', as at Mogadishu, a ploy in which the assault team breaks inside through two or three points, hitting the terrorists from a distance, the SAS and GIGN also practise flooding the aircraft through as many points as possible. Some men might try to gain access through the luggage compartment beneath the fuselage or roof hatches to try to get in as near to the terrorists as possible and engage them in close quarter battle. Taking the aircraft from many points, however, requires meticulous, split-second coordination, otherwise it can go disastrously wrong.

When Egyptian commandos of the counter-terrorist 777 force tried storming an Egyptair flight hijacked to Malta by having part of the team enter through the roof hatch, they overcharged the explosives placed there and the plane caught fire. Since none of the exits had yet been opened, ladders were not in place and commandos crouching along the wings ready to blow open the emergency doors were jolted off the plane by the violent explosion. Fifty-seven passengers died trying to get out of the burning aircraft. One of the terrorists escaped.

Counter-hijacking involves assessing everything about an aircraft – its weight, how much fuel it has, the number of people on board, the quantity of luggage. It also requires training a normal-sized team of about twenty men to move inconspicuously through the security checks of an airport in civilian dress while carrying concealed weapons. The ability to move as anonymously as the terrorists themselves became very necessary for Delta Force when it was called in to deal with the hijacking of a TWA flight in October 1985. The flight from Athens had been diverted to Beirut by two gunmen of the

Islamic Jihad. Some hostages were released in Lebanon as ten terrorist reinforcements were picked up and the hijacked airliner flew on to Algiers. The Algerian government was highly ambivalent about allowing a US military rescue mission on its territory but let it be known that if the American team acted with sufficient speed and discretion, it would turn a blind eye.

The Delta team were inside the airport at Algiers, prepared to assault the hijacked plane by the time US Secretary of State George Shultz authorized the mission to go ahead. But mid-level officers in the State Department's 'crisis management team' held up the paperwork and the hijacked flight took off to return to Beirut, where the 32 hostages were taken off the plane, distributed around Shiite-held areas of the city and a rescue operation was no longer feasible. The U.S. was forced to cave in to the terrorist demands, pressuring Israel to release 700 Shiite prisoners. The carefully planned shuttle between Algiers and Beirut was a clear indication that the terrorists had also been learning about special forces methods and were careful not to remain in any single airport long enough to allow a counter-hijack operation to be set up.

By their nature international incidents, aircraft hijackings require different counter-terrorist teams to work closely together and become familiar with each other's methods. During incidents in which a hijacked aircraft moves between different airports, a constant information flow is required between the special forces concerned as the jurisdictions change. During one hijacking by Abu Nidal in which the plane landed in Cyprus before going on to Algeria, the SAS and GIGN coordinated their activity and information at all times.

'While the plane was in Cyprus the responsibility for a rescue operation was with the SAS and when it flew on to Algeria, where we had better contacts, it switched over to the GIGN,' explains Philippe Legorjus. 'We had to get all the up-to-the-minute information the British were gathering about the plane, passengers, terrorist team, and know the entry plan they had devised as we planned our own.' Both units set up a computer link to interface their information instantly. CT organizations cooperated in a similar way during another hijacking between the former French African colony of Niger and the former British colony of Nigeria and since there were West Germans on board the plane, the GSG9 got involved as well.

When 'engaging down fuselage' in storming an airliner, a counter-hijack team needs to move quickly down the aisle in a single file, shooting at the targets over each other's heads. A 'continuous fire' drill developed by the GSG9 and adopted by other CT teams to get themselves used to firing in support of each other within very close proximity consists of three lines of men moving on static targets. Each line alternates between standing, kneeling and lying as they move past each other firing their sub-machine-guns until all the men are within point-blank range of the targets. It is part of the instinctive shooting techniques which any counter-terrorist operative masters when he enters the 'killing house'.

The German-made Heckler & Koch MP-5 is the universally favoured sub-machine-gun for counter-terrorist work. Firing ten rounds per second effective at 200 metres, the MP-5 has a carefully engineered bolt which rotates instead of moving back each time a fresh round enters the chamber. This allows for greater accuracy, avoiding the tendency of other sub-machine-guns for the rounds to climb upwards as they are fired. For this reason the MP-5 is the most perfect weapon there is for indoor close quarter battle. The handguns used by the different units, however, tend to vary in relation to differing methods and the focus of the mission. The GIGN mainly shoots with .357 magnum revolvers, while the SAS and GSG9 use 9mm automatic pistols. Delta prefers the old US Army standard Colt .45 because its bigger round travels at subsonic speed and is less likely to traverse the body of a terrorist and wound a hostage as could easily happen in the cramped space of, say, an aircraft cabin.

The GIGN's preference for the slow, high-powered revolver over the racy automatic is conditioned by its doctrine of controlled force. Being a police unit called on to intervene in both terrorist and purely criminal cases, the French group requires every round to guarantee strong stopping power while avoiding unnecessary casualties. The magnum's powerful bullet and the accuracy of its long barrel provide that assurance. Prouteau, the former head of the GIGN, even developed a miniaturized telescopic site to be used with a magnum to engage hostage takers from a distance. The emphasis in CQB training for the SAS, on the other hand, reflects the extreme environment of guerrilla war in Northern Ireland, where deadly force is required, along with instant speed and flexibility, in circumstances of close proximity during ambushes against multiple groups of heavily armed, well-trained and dangerous terrorists.

The GIGN and the GSG9 mainly practise shooting down range, taking carefully aimed shots at targets from distances of twenty-five metres. The SAS get up close to point-blank range, firing instinctively as they squeeze off 'double taps' from their 'nine milli'. The course starts off on a dry run with a big rugby-player-type rushing the trainee, who has to draw the unloaded gun from a holster at his hip. If it's not out and pointing in the fraction of a second necessary to stop the charging man, the trainee is bulldozed.

'Small blokes like you always draw faster,' was the backhanded compliment I got from the instructor. 'You've got that additional motivation.' At a crouch you punch out the pistol in a fast aggressive move, like a boxer going for a straight jab. The left hand closes over the right one, which firmly grips the weapon. Aiming the gun, not the sights, you squeeze the trigger in two instant shots at a cardboard bull's-eye centred on a picture of a man with a revolver which moves towards you on rails. You should be down to the last bullet of your thirteen-round clip by the time the perforated figure is right on top of you. All your shots should be grouped closely together. 'Not bad,' says the instructor. 'You grouped six shots right in the bull's-eye but you have a tendency to shoot low and to the right. Try correcting that by aiming further up and towards the left.' The conveyor noise begins again.

The drill goes on for two hours, then four hours at a time. It can last six, even ten, hours a day between breaks. Fast drawing results in accidents even among experienced specialists like one Delta man with an exemplary special forces record who shot himself in the leg when his finger contacted the trigger too quickly as he pulled out a Colt .45. 'He recovered quickly but his ego took a longer period to heal,' remarks a fellow member of Delta. Sometimes trigger mechanisms are so well greased that shots squeeze off prematurely under the slightest pressure.

The fat, shaped 9mm bullets seem to dribble around you as clip after clip of ammunition gets packed until the coil is tightly compressed under those thirteen rounds and is snapped into the magazine inside the pistol butt. The metallic scraping noise of the cocking handle being pulled back and then released has an ominous quality as the safety-catch is released; it clicks into position and the first round enters the chamber. Eyes get irritated from the fumes mixed with the smell of cordite and gun grease. Headaches and even nausea occur.

From a boxer stance you learn to manoeuvre into a weaver stance as targets begin to move laterally as well as towards you and away from you. 'Centre on your asshole, like you were taking a crap,' advises the instructor. 'The asshole is very important. Pivot on it and it will move your shots right and left. Legs control you up and down but the asshole is your axis.' You then drop to your knees and are next shooting in the lying-down position called 'the banana' as targets pop up from everywhere.

There are different aiming techniques, such as pointing with the index finger placed like a rudder along the side of the gun to direct the shots while the trigger is squeezed by the middle finger. The instructor contrives problems like loading empty shell casings into the ammunition clip to see what you do when the round doesn't fire. 'That's good,' he says under his breath as you eject, go on to the next and keep firing; he is satisfied that the trainee reacts properly in a 'pressured situation'. Index cards are pinned to different parts of the body on the cardboard dummy, conditioning the shooter to group his shots on varying areas of the target. The favoured aiming point is on the bridge of the nose, to inflict an instantly fatal head wound. The shooter becomes acquainted with different types of munitions, including silent bullets with rubber coating on the lip of the cartridge to absorb the impact of the firing pin. The bullet loses speed but is still very effective at close range.

During detached, clinical classroom instruction, you become familiar with all types of automatic pistols used by the SAS: 9mm Brownings as well as Berettas, Sauyers, Walther PKKs, the German-made PS9, and small .22 pistols and .38 revolvers, both of which are easy to conceal and useful in undercover missions. Guns become a universe of their own and by the time an expertly trained shooter is prepared for action, he should be able to assemble and disassemble the parts of any of them blindfolded.

When you can draw your gun and fire accurately within a fraction of a second, you enter the next phase of CQB training – selective shooting. As you walk into an indoor range you see three life-sized picture targets staring at you: a mean-looking guy with a gun, a blonde temptress holding a hair-drier and another man with a camera. You have no problem in drawing your weapon and hitting the 'bad guy'. In the next line-up you see the same three pictures but with the objects switched around on the small rectangle in the middle. The mean-looking guy now wears a policeman's badge, the blonde holds the gun and the other man has the hair-drier. Instinctively you hesitate as the mind reassesses the threat, which is now posed by the woman. If you automatically shoot one of the 'good guys' you lose points but if you become confused and hesitate for over a second you are out. After intensive four-hour sessions, an instinctive shooter will develop the capacity to identify the threat within the half-second draw and firing speed he has already developed.

The number of targets is increased gradually from three to six. In one line-up they all have guns and the next time around they all carry hair-driers and cameras. Situations are further changed so as to totally condition the mind to respond to threats. If one of the targets has a knife, you shoot it. The SAS pops up the image of a priest holding a cross like a sub-machine-gun. You get marked down if you hit that one. The GSG9 uses a skeet range to practise selective shooting. As clay pigeons painted in five different colours are hurled into the air the order comes to fire just on blue ones, then on blue and red, and so on.

As part of his final exam, the special forces gunman walks down a long, dark alley with his pistol. A door will suddenly slide open and he pivots around to see a woman holding a baby. He holds his fire and the door slides shut. In the next two seconds a door opens on the other side to reveal two men holding sub-machine-guns. He shoots double taps into both in less than a second. Then, as he continues down the alley, another entrance blows open, with one man standing inside with a revolver holding another one who is unarmed. This goes on for forty-five minutes and his speed, reflexes and accuracy are marked throughout. The exercise is repeated with a sub-machine gun.

'It's a question of getting the mind to control a man's reflexes in the highly pressured situation of a room clearing or an ambush,' explains an SAS instructor. 'The fear factor, the released anticipation, the sudden rush of adrenalin, the sweaty hands, the excitement mixed with panic, can easily lead to mistakes during moments in which well over one hundred rounds are being expended in a matter of seconds.'

In the full killing house exercise, one group abseils down from a flying helicopter on to the roof while another dozen men take up crouching positions against the walls on either side of the main entrance. Shotguns, sub-machine-guns and pistols are held at the ready with safety-catches off as the demolitions expert steps up to the front of the door, tapes a frame charge around the steel lock then moves back a few steps and pulses the button on the detonator cord. As it blows open the door, the few grams of plastic give off a shock wave and an ear-splitting bang. Three men instantly rush in through the smoke, flames and flying debris. The one running in front may carry only a pistol while the other two cover him with their MP-5s. Their body armour is lighter than that of the men following behind, who are more likely to take a hit from the enemy, whose effective reaction time is calculated to be a point three seconds after penetration. As the assault team floods into the ground floor, the advance element starts bounding up the stairway, covered by the last man, who has entered with a repeater shotgun. He wears the heaviest body shield, protected with layers of ceramic plates against high-velocity rifle rounds. He is in the most exposed position.

Clearing individual rooms through a floor, five-man teams work down each corridor. One man pumps out slugs from a Remington shotgun, shooting out the door locks, while the other four men divide into pairs and take each room on either side of the hallway. The SAS automatically tosses in flashbangs on entering. The GIGN and other CT units, fearful of the fire hazard, don't resort to stun grenades unless they are considered necessary. The pair rushes through the door simultaneously, going through a pre-choreographed ballet. The main movements are either (a) 'slingshot', in which they move laterally from the entrance to the two near corners, firing double or triple MP-5 bursts at identified 'hostiles' or (b) 'cross-over', in which they criss-cross each other to control the enclosed space from opposite walls. The second method, which may be necessary if several terrorists are trying to take cover behind hostages, is the riskiest option since the pair may end up shooting in each other's direction. Each room has to be cleared in four seconds and the entire floor should take no longer than fifteen seconds. As in most other special forces realistic exercises, live ammunition is used.

It used to be SAS practice that in a CT final exam, the instructors would sit among the targets, role-playing hostages while the assault pair stormed in, shooting up the terrorist dummies with their MP-5s. But an accident eventually happened and an instructor got killed. Delta tried to solve the problem of the fatal dangers posed in realistic CQB training by making wax bullets. A regular assembly line was set up which involved using a kinetic bullet puller to remove the bullet, empty the powder and dip the round into hot wax and then reassemble the round. The wax bullets might induce trauma but wouldn't kill anybody. But then virtual reality came to CT.

An SAS man storming through a killing house today avoids hitting the hostage diving for cover under furniture while squeezing his triple burst at the terrorist who comes at him shooting a revolver. A single shot has to be carefully aimed through the MP-5's target acquisition beam, by centring a red dot right on the forehead of another gunman grabbing a hostage to shield himself. The action taking place is the result of interfacing three-dimensional imagery through bulletproof walls. The room-clearing team is in one room watching the reaction of the terrorists and hostages in the adjoining room and vice versa. The scenes in the different rooms are simultaneously filmed and projected through closed-circuit TV cameras on wraparound screens. The movie stops when the exercise stops and either all terrorists – or the CT team – are neutralized.

CT units lacking the budget for such expensive training aids use life-sized pre-filmed videos projected on to the back wall of a killing room. 'But the problem with videos,' explains Christian Prouteau, 'is that the men can memorize the movie.' The GIGN prefers to project high-definition colour slides, which it changes during brief intermissions between a team's multiple assaults. 'The action should be so quick as to prevent very much movement inside of the room in any case,' thinks Prouteau. 'What the team needs to become accustomed to is reacting to the different situations which they may encounter.'

Just as a football team studies their various moves after a game, the SAS and other CT units run slow-motion replays of their rehearsals. 'You were slowing up, Mike,' the instructor says, freezing the video at the moment that it shows the trooper hesitating as a terrorist springs at him from the inside of a closet. 'You almost lost a half second before slotting him.' The video stopped altogether during one room-clearing practice in the Australian killing village which was being closely observed from outside the door by a high-ranking general when a flashbang exploded prematurely in the hand of the shotgunner. While the assault pair in their fire-retardant suits had already run inside the room, everyone along the hallway was sprayed with the burning phosphorus. The shotgunner jumped up screaming, while the general looked down on his singed body hair exposed through the gaping hole burned in the belly of his dress shirt.

In no other type of military training is as much small ammunition expended as in CT. In one year the Australian SAS consumed the Australian Army's entire stock of 9mm ammunition. The few hundred men in the US Navy's SEAL Team Six get a bigger annual allowance of bullets for live-fire training than the entire US Marine Corps. The expensive but delicately engineered Mauser PSG-1 rifles are fired so much by SAS snipers that the bores of the barrels get burned out and replacements have to be constantly ordered. The mechanisms of pistols which are now subject to stress tests of 10,000 rounds of continuous firing have to be customized by gunsmiths to withstand the pressures of CT training in which an SAS man fires an average of about 5000 rounds per week.

'Many of the men also end up suffering from post-traumatic stress disorder,' says Bob Mawkes, citing this as the reason why the SAS rotates all squadrons through CT duty in six-month shifts. 'It gives all the men a chance to train at it and stops anyone from getting too burned out. CT gets very frustrating.' In Australia the duty squadron gets regularly stood up for the Commonwealth Games once a year. But between the endless killing house exercises the men have to live in a netherworld of false alarms. Being on CT duty may give them a comfortable period of living at home but it is not uncommon to get beeped out of bed in the middle of the night for an emergency.

During a quick briefing at headquarters they are convincingly told that a terrorist team has been spotted at a certain airport preparing a hijack. They zip into their black, fire-retardant suits, pack all their equipment, ammunition, stun grenades and weapons into holdalls, board the waiting vehicles or helicopters only to arrive at the scene and be told that it's back home again. It was only a readiness exercise. Squadrons in 'war roles' will generally know if and where there is a real war. Those on CT have no such certainty. The whole point about terrorism is that it can happen anywhere at any time and the counter-terrorist's job should be, ideally, to pre-empt the incident before it makes the news headlines.

'We are a unit for war within peace,' is how Christian Prouteau describes the CT mission over a lunch of sole meuniere and white wine at a restaurant off the Champs-Elysées. Working now as a security adviser to President Mitterand, he is sitting with top executives from Air France and other large corporations. He explains that as a police unit, the GIGN gets plenty of realistic practice in dealing with siege situations involving ordinary bank robbers, lone psychotics, prison mutinies and a variety of relatively mundane criminal episodes. Some are not quite so common, however; like the time when the group was called out to rescue a civil servant being held at gunpoint by a dominatrix whom he frequented in his lunch hour.

Prouteau once had a desperado 'make his day'. Outperforming Clint Eastwood's Dirty Harry, the GIGN's founder pulled back the firing-pin with his right thumb and released it with one hollow-sounding blast from his magnum revolver, driving a high-powered .357 bullet right into the shoulder of a bank robber holding a gun on a little girl. It's the method favoured by the GIGN. 'Hitting the shoulder of the arm holding the weapon eliminates any threat a man can pose,' Prouteau explains. Shooting to disable rather than kill is the GIGN doctrine for most hostage situations, even when it involves terrorists. 'The decision to take a life depends on the situation. We prefer to neutralize with a single shot in close quarter battle and take the terrorists alive,' says Prouteau, who took the decision to use lethal force on his own initiative when there was no other option during the school bus episode in Djibouti. GIGN's current commander, Denis de Favier, adds, 'If hitting the shoulder cannot work then we aim a single shot to the head. It's easier to discipline this with the revolver than with automatic weapons.'

'Military units like the SAS treat hostages as a military objective and terrorists as enemy soldiers,' says Prouteau, who had many discussions on the subject with his British opposite number, Mike Rose. Instructors in the SAS and the SEALs are the first to agree that 'once we go into a siege situation it means the enemy gets killed. Our doctrine is one of total aggression'. Rose appreciates the GIGN's concern with sparing life but adds that 'it would be very difficult to teach my soldiers that the life of an enemy in front of them with a gun is more important than their own'. He also objects to the 'authoritarian' method in some units, citing Germany's GSG9 as an example, in which rules, tactics and plans are decided at the top by a 'technical group' which passes them down to the men who have to perform the operation. He makes heavy gestures with his hands as if mimicking some cumbersome unwieldy machinery. 'A plan should develop from the bottom upwards and be approved in close consultation with the men who have to execute it.'

There is a relaxed atmosphere inside the headquarters of a counter-terrorist unit. Officers and non-commissioned ranks address each other by first names. Salutes and other military formalities are dispensed with. Although regulation army camouflage is usually worn in the SAS, where bleached, weathered fatigues denote status and seniority, members of the GIGN and other CT units sit at their desks and walk around in trainers, work-out clothes and vinyl ski suits in purple and other luminescent colours. Looking well-scrubbed, lean and fit, they could all be the prize jocks of some athletic team in the clinical setting of whitewashed walls and polished linoleum floors.

Large windows overlook the surrounding army compound outside Versailles where hapless conscripts stand on parade uniformed in green olive drab. Plaques and trophies from innumerable shooting, abseiling and martial-arts competitions adorn the spartan room. A blown-up team photograph which appeared on the front cover of *L'Express* stands on the first landing past the entrance hall, showing a rifle-toting group of the GIGN with black-smeared faces and green khaki combat gear who put down a rebellion in the French South Pacific colony of New Caledonia.

CT jocks come in and out of the gym, where they swim laps in the twenty-five-metre indoor pool, do weight training and practise karate and judo. Prouteau could break five bricks with a single karate chop. Black-belt proficiency is currently maintained by thirty members of the GIGN. Captain Denis de Favier, who, with his wet combed curly blonde hair, looks like he has just come off the running track, explains that although black-belt level is not required, his Gendarmes must be capable of mastering the self-defence and unarmed combat techniques peculiar to CT, such as quickly blocking and disarming an armed gunman while weighed down with body armour and other assault equipment. We go for a mid-morning *demi-tasse* in a small bar, crossing a garage holding a fleet of the GIGN's blue armour-plated radio cars and vans. Everyone is clean-shaven, with their hair still wet and towels hanging around their necks, as if they've just stepped out of a shower and sauna.

The sudden eruption of demolitions charges as a group outside practises entry techniques, the crack of pistols and bursts of sub-machine-gun fire, are the final reminder that this is not the Washington Redskins or Manchester United. Cars screech around into 'bootleg curves' as drivers accelerate while simultaneously applying the hand brake to achieve a high-speed U-turn. This is a standard CT technique to evade a street ambush. Should the assailants try to block the escape by boxing in the target with another cut-off vehicle driving across the back, the solution is a 'controlled crash' where the driver steers his vehicle to impact on the front corner of the blocking car, spinning it sufficiently to open an escape path. So many cars turn over and are written off or damaged that Delta, among other units, has often resorted to renting them. Otherwise, the normal Army procurement system could never keep up with the demand for practice vehicles also used to train men in surveillance, counter-surveillance, pursuit, protection and rescue.

CT units practise setting their own street ambushes. One group of about thirty men prepares to assault a minibus in which terrorists are holding hostages. They could be on their way to board a plane as Black September did in Munich and Salim wanted to do at Princes Gate. Various ruses such as contrived accidents could be used to bring the vehicle to a halt. The SAS has devised a method of blowing out all four tyres simultaneously, a secret which they share with no one.

The instant in which the bus screeches to a stop, about a dozen men run up carrying stepladders, and form an L along its side and back. Crouching out of view from the inside, they place the ladders beneath the windows for the next wave, who immediately rush forward, mounting the stepladders in one swift bound and breaking open the windows with their pistols. Pointing them inside, they cover the second assault group, who are forcing their way in through the side doors. It all happens in five seconds. Delta practises bus take-downs with role-players inside the vehicle. In one exercise a woman hostage jumped up in panic as the assault team stormed into the vehicle. Shot in the face with a wax bullet, she suffered severe trauma.

The Grupo Especial de Operaciones (GEO) of Spain's National Police might be considered by some as a poor cousin of the CT community – no fancy killing houses, no flashbangs, nor even any proper targets for its snipers, who often use saucers. Demolitions and room-clearing exercises take place in a small brick hut in the dusty courtyard of the castle-like compound which rises from the plains of Guadalajara, about an hour's drive from Madrid. The GEO was founded in 1980 by the first democratic government to succeed the forty-year dictatorship of General Franco. Its organization and training are modelled on Germany's GSG9.

When the US Navy's SEAL Team Six came over for joint exercises with the GEO in preparation for a possible 'Munich' at the 1992 Summer Olympics in Barcelona, the Americans were amazed to find that their Spanish counterparts did not use underwater rebreathers. However, according to its director of operations, Inspector Carrion, the GEO has more experience of hostage rescues and the hard arrest of terrorists than many other special forces units.

The GEO is on constant alert against the ongoing threat of the Basque separatist guerrillas of ETA, which rivals the IRA as one of the most dangerous, ruthless and tenacious terrorist networks in Europe, with international connections stretching from Libya to Cuba. ETA has financed itself to the tune of two billion dollars through a highly efficient kidnapping and extortion racket enforced among the business community in northern Spain. Drawing its model from such Latin-American urban guerrilla movements as the Tupa-maros, who made kidnappings a speciality, ETA, through its ability to raise huge funds and generate publicity and political instability through multiple abductions, has kept the struggling GEO on tenter-hooks.

Comandante Holgado had taken advantage of the Saint's Day festivities in the Basque village of Trasmoz to blend in with the crowds. The GEO commander had brought along a policewoman to make it look like they were a courting couple. Holding hands, they strolled around a grey stone house with rows of small windows, indistinguishable from the other dwellings in this community of 175 inhabitants. As in many other mini-municipalities nestling in the Basque Pyrenees, ETA's political front, Herri Batasuna, the People's Party, controlled the town council, knowing everything that went on in the village, particularly who came and who went.

The environment seriously complicated a police surveillance on the building, in which, GEO intelligence indicated, ETA was keeping a hostage. Any unusual presence or activity around the village ran the serious risk of tipping off the two or three terrorists known to be inside the house guarding any one of several recent victims of ETA kidnappings, including the father of the world-famous singer Julio Iglesias.

Holgado and his companion lingered by the stone wall in front of the entrance. He pulled her close as he peripherally examined the main door, calculating its size and thickness. It was a double gate made of heavy wood. He went through some mental calculations of the quantity of explosives which would be needed to knock it down without causing major damage. Some thirty grams of the Spanish-made plastic explosive Goma Dos should do the trick.

Arm in arm, they strolled around the side of the house, where Holgado suddenly embraced the woman as he distinctly saw that one of the windows on the second floor was blocked from the inside.

The thirty-man GEO assault team driving up the narrow two-lane road to Trasmoz that night in blacked-out vehicles with their headlights off still did not know if it was Iglesias or a Basque industrialist by the name of Lippenheide that they were going to rescue. The cars and vans stopped about two kilometres outside the town and the men, wearing trainers and plain clothes, piled out into the cold, rainy darkness. After strapping on their body armour and Kevlar helmets, they made all the last-minute checks on their demolitions equipment and German-made armament – Heckler & Koch MP-5s, PS9 pistols and Mauser PSG-1s on which the snipers fitted infrared telescopic sights.

The men could hear the last fireworks of that day's fiesta crackling in the distance as they crept up on the village in single file through the fog and drizzle. It was just after 2 a.m. when snipers and cut-off groups took up positions around the house and the main assault force of just over a dozen men stealthily drew up to the stone wall. Pistols and sub-machine-guns were aimed over the wall as the bearded Miguel Angel, an NCO, crawled up to the main door, taped on the frame charge, stepped off to one side and pulsed the detonator cord three seconds later.

The explosion shook the house as the thick wooden gate was blasted into flaming splinters. Angel led the way in, storming through the debris and mounting the narrow flight of stairs to the second floor. Doors came crashing down into all the bedrooms. Caught in their beds, both stunned ETA gunmen and one woman had barely time to think about reaching for their weapons before they were staring into the barrels of GEO sub-machine-guns.

It was Iglesias that Angel found when he blasted down the door of the small bedroom at the end of the corridor. 'He looked weak, sitting up on his bed. Understandably shocked, he didn't say very much at first.' A cupboard was chained to the window and the captive's only other furnishing was a pot to piss in. Without a single shot fired, the GEO had liberated ETA's prize hostage. It was Spain's answer to Princes Gate.

In a sniper action which could serve as a textbook example of selective shooting, the GEO once ambushed an ETA 'commando', making a night-time landing on a rocky, windswept beach near the northern port of Pajares. The police had precise intelligence about the landing point of the ETA cell entering Spain by boat from the south of France. The terrorists planned to rendezvous with a car along a road running parallel to the beach. Their mission involved the kidnapping of one of the richest industrialists in northern Spain.

With the military professionalism of a SEAL team, the five Etarras manoeuvred through the strong surf in a motor-powered inflatable Zodiac boat and washed on to the shore in the dark, moonless night. After collapsing their landing-craft and hiding it among rocks, they moved inland towards a coastal road which ended at one of the water breakers at the harbour's entrance. They were spread apart, armed with sub-machine-guns and grenades. 'Freeze! Police!' came the command from GEO officers emerging in their windcheaters and watch caps from behind rocks, shining torches on the terrorists, who responded with a barrage of sub-machine-gun fire. As the arresting team ducked for cover from the spray of bullets, snipers sighting with their infrared scopes from positions about 100 metres away opened up with Mauser PSG-1s. Four terrorists were instantly shot dead. A fifth who threw down his weapon and raised his arms was taken unscathed.

In a cavernous indoor range, GIGN snipers do a morning warm-up shoot. Lying on an elevated gallery with their custom-designed FRF-1 7.62mm high-powered rifles resting on bipods which fold out of the stock, we are hitting targets measuring 15cm by 15cm at a distance of 200 metres, trying to group our shots into the 3cm bull's-eye. It's barely distinguishable, even with 20/20 vision. You have to aim almost instinctively for the centre of the concentric circles, which appears like a small black dot between the two converging horizontal lines and one perpendicular vertical stripe of the GIGN's distinctive telescopic sight.

The rifle's heavy recoil forces you to re-aim as you thrust the manually operated bolt backwards and forwards, releasing the burnt casing and pushing it shut as a new round enters the chamber. After emptying the twelve-round magazine, we all walk over to inspect the targets. Most of us have grouped about half a dozen shots into and around the bull's eye. One of the snipers got them all into the bull's-eye, which he has turned into one big hole.

Then the targets are moved to 300 metres and the snipers raise their sights accordingly, turning the windage drum on the side of their telescopes to calibrate their vision in relation to the rise of the ammunition. GIGN shooters are expected to fire accurately at distances of 600–800 metres. The specialized snipers should be able to hit at a range of 1000 metres with the heavy .50-calibre armour-piercing rifle.

A specialized sniper normally practises on his own. Aiming his rifle over an empty ten-foot frontage at his 'normal stalk' of 300 metres, Sergeant Mick McKintyre suddenly sights a 4cm-diameter target popping up. He has only three seconds in which to get it on the pinpoint dot between the horizontal cross-hairs of his SAS sights and fire. Four more targets will appear at erratic intervals in the next ten minutes. Shooting in an outdoor range, McKintyre, who became an SAS sniping instructor, also has to assess the atmospheric conditions which may affect his accuracy. In hot, damp climates shots tend to rise higher while the thin air of Arctic temperatures makes them drop. He feels a gust of wind coming from a southerly direction which may blow the next round just a bit off course. Without taking his right eye off his sights, he turns the drum on the side of the telescope just one click.

A new sniper weapon is being introduced whose target-acquisition laser beam, which appears as a small red dot in the centre of the telescopic sight, blocks the trigger release until it has acquired an accurate line of fire. With a range of several miles, the light-calibre 22 rifle, used with a silencer, is 'the ideal assassin's weapon' according to a German GSG9 sniping instructor. There is no risk of an inaccurate round warning off the target before the sniper corrects his bearings. The atmospheric conditions which may affect the bullet's trajectory are also factored in automatically by the rifle's computer. However, GIGN leader Captain Denis de Favier dismisses it for counter-terrorism. 'We cannot allow the decision to shoot to be made by a machine. When intervening in a terrorist situation, the decision to fire has to be a human decision based very much on the circumstances.'

When GIGN snipers stalked the leaders of the nationalist Kanak movement to their hideout through the equatorial jungle of New Caledonia in 1985, 'the decision to fire was very quick', says Philippe Legorjus, who commanded the operation. It was 'a political decision' which had already been taken by the French government, who wanted to put an end to the serious disturbances which had broken out in its island colony straddling the Pacific and Indian oceans. There were hostages, riots and the local barracks of the Gendarmerie had been attacked. Re-establishing order was imperative, by all means necessary.

The four snipers took up positions 250 metres in front of a small farmhouse in the middle of a jungle clearing. After lifting the Plexiglas visors over their Kevlar helmets, they took aim through their night-vision infrared telescopic sights. The Kanak leader, Machoro, was sitting on the floor of the small entrance porch, flanked by two other men, one of whom was lying down. Aware that they were being followed, the men nervously observed their surroundings as they talked. Two sentries armed with shotguns patrolled nearby.

Three seconds after the snipers had sighted down on the front porch from ninety to eighty-degree angles and reported having acquired their targets, the order came back over their walkie-talkies to open fire. The marksman positioned furthest to the left immediately shot Machoro and the other Kanak sitting next to him, killing both men instantly. The next sniper, firing at a straight angle, took out the man lying down and the sentry in front. As the other guard appeared at the side, pointing his shotgun, he was immediately hit by several rounds fired from the right.

It was a French special forces hit job. The fourteen-man GIGN team dispatched to New Caledonia was backed up by elements of the Foreign Legion's 2 REP, who were sent uniformed as Gendarmes. The broad-shouldered Legorjus also liberated a group of hostages being held by the Kanaks inside a cave. 'We believe in order for the right reasons,' says the one-time student radical who left the GIGN shortly after the New Caledonia operation.

No backup snipers with high-powered telescopic rifles were positioned around Gibraltar on 6 March 1988 as a four-man SAS patrol closely followed three IRA terrorists who had just parked a vehicle suspected of containing a bomb in front of Government House. Ambiguous orders, confused intelligence, overlapping jurisdictions and other factors affecting the somewhat disjointed efforts by British and Spanish security services to prevent terrorist carnage erupting at the Colony's ritual Changing of the Guard ceremony, would have obstructed a clean stand-off job – if, as many believe, the objective of Operation Flavius was 'shoot to kill'.

The stocky, bull-necked Sergeant Bill, whose 9mm automatic stuck out of the backside of his jeans with two spare clips showing out of each back pocket, was hiding along a line of bushes as he observed Sean Savage, Mairead Farrell and Daniel McCann walking back towards the Spanish border. The thuggish-looking Para, with his sand-papered hair, pencil-thin red moustache and muscles bulging out of a tight T-shirt, looked somewhat out of place among the

toddlers and Moroccan nannies who crowded the children's playground on that warm, sunny Mediterranean afternoon. But things
were moving so fast that the SAS hadn't had time to arrange for a
more appropriate OP. 'It was the only place to hide. It was all rather
improvised.'

According to most accounts, either Spanish police surveillance of
the terrorists, who had been watched for days since their arrival at
Málaga airport, had broken down during their border crossing or
there had been a failure to communicate it to the combined Army,
Gibraltar Police and MI5 control team assembled on the Rock. But
the presence of the three identified IRA terrorists and the hired
Renault car being driven by Savage was not detected until they
were all practically gathered outside Government House, a short
distance from the frontier, as most things are on the Rock.

The SAS team had been briefed to expect to be facing some of 'the
most dangerous terrorists in the IRA', the worst that they would ever
encounter. At the time the guerrilla organization was preparing a
terrorist bombing campaign against British military bases overseas.
The track records of each member of the IRA team had been read to
them in detail, including that of Mairead Farrell, who had already
served a ten-year prison sentence in connection with a deadly car
bombing in Belfast. In the words of one SAS sergeant, 'we were so
wound up that if a chick had farted we would have fired'. When Sean
Savage split away from Farrell and McCann, exchanging newspapers
outside a Shell petrol station, and walked in the opposite direction up
Winston Churchill Avenue towards the intersection with the Land-
sport Tunnel, 'it looked like some sort of signal'. As Savage bumped
into one of the SAS men following them down the street, tensions hit
the point of no return. The adrenalin pumped through the tensed-up
bodies and in the next few seconds so did bullets.

Sergeant Bill ran out of the playground, crashing into a strolling
couple with a baby carriage. 'Sorry, mate,' said Bill, excusing himself
as he leaped to join the fourth member of his patrol, a man with long
blonde hair and wearing a denim jacket who was getting into position
at the street corner. Bill looked back at a police van parked in the side-
street across from the playground. Its siren began to wail at that
instant, catching the attention of McCann, half a block away, who
stared backwards and straight into the eyes of the SAS man now just
some feet away and closing on him. Since the terrorists were probably
armed and possibly holding a remote-controlled detonator for what

could be the car bomb parked outside Government House, the mere feeling of having been recognized was enough for instinctive shooting to take over.

Whether McCann held a weapon hardly seemed to matter. The individual himself, now aware that he was being followed, presented a threat. The mere eyeball contact and sense of recognition of a dangerous terrorist whose bloody past was as well known as his current intention to slaughter children and tourists enjoying a public display of military pageantry unleashed all the training which had been instilled into the SAS gunman over months of immersion in the killing house.

Soldier A went into his CQB crouch as he pulled the 9mm handgun, with its cocked firing-pin, from his waistband, simultaneously clicking off the safety-catch with his thumb. He started shooting in the next half-second, just as he had been trained to do, squeezing out double tap after double tap, getting to point-blank range and not stopping until he had expended the thirteen-round clip.

Soldier B attacked Farrell in the same way and Sergeant Bill manoeuvred with the blonde trooper in denim to form a pincer on Savage, firing on him from two angles before the terrorist could make a dash for the darkness of the tunnel carved out of Gibraltar's old white-granite fortress walls.

What to the untrained eye of a perfectly rational bystander might seem like pathologically sadistic overkill was to the SAS man schooled in the efficient administration of death the true-and-tested method of assuring that an enemy would have no possibility of reacting with a hidden gun or grenade or a button operating a remote-controlled bomb.

'Initially, I aimed at the centre of Savage's body,' says Soldier C, who claims to have shouted orders to stop as he drew and shot his gun. 'I fired. I kept on firing until I was sure that he had gone down and was no longer a threat initiating that device. I fired the first round and carried on firing, very rapid, right into Savage's body as he was turning and fell away to the ground. The last two rounds were aimed at his head. [These passed through Savage's head and made a crater measuring three by four inches in the pavement.] This was possibly inches away from the ground, just before he became still. I kept on firing until he lay still on the ground and his hands were away from his body.'

8

WHERE THE SHARKS SLEEP

'Fuck you all, cockbreaths!' is how Richard Marcinko began his speech to the first assembled group of Navy officers and seamen he had personally selected for SEAL Team Six. It was 1980. He was late in the game and the bearded naval commander stood in front of a giant American flag. 'You know what we are all here to do – counter-terrorism. And what does counter-terrorism mean? It means that we will fucking do it to them before they fucking do it to us.'

SEALs who have sat through Marcinko's innumerable harangues describe how he drank down shots of bourbon and beer chasers during his 'shit dances'. But the creator of SEAL Team Six was expressing the very ethos of military CT. SEALs had been running pre-emptive hits against guerrillas since Vietnam's Phoenix Program, even before the SAS did it on a regular basis, and the new counter-terrorist mission would be a particularly aggressive and dangerous one.

Swimming on to an enemy-controlled offshore target is a strenuous physical feat to accomplish in the best of circumstances. To suggest to a commando swimmer who has just clambered on to an oil drilling platform after releasing from the escape hatch of a submarine in a rough ocean that he should shoot to the shoulder of a terrorist would be inadvisable. The GIGN, which started training in water operations when the ex-combat swimmer Philippe Legorjus took command in the 1980s and is the only French special forces unit allowed to practise on cross-Channel ferries, concedes that in a maritime environment, normal rules of engagement cease to apply.

'You are the system, gentlemen,' Marcinko would tell his men. 'The buck stops with each and every one of you. I've bought you the best gear that money can buy and you will take care of it. If your equipment fails it's because you've fucking failed. So I will not accept any goddamn excuses ... Failure is on your shoulders.'

'The sea is an unforgiving place so you have to do everything right,' says Ram Seagar, who developed the CT role for Britain's Special Boat Service (SBS) when North Sea oil exploration began in the late 1970s. During one of his first missions in Borneo, Seagar missed his rendezvous with a submarine when swimming back from a reconnaissance on an Indonesian beach. 'We will have to keep swimming until we get to Singapore,' was the only answer he could give his other trooper as they drifted amid the ocean swells. 'He went apeshit,' Seagar recalls. The sub finally found them.

From the moment in which a combat swimmer makes his first fifteen-metre vertical dive underwater and starts learning to use rebreathers at shallow depths, he must attune himself to an alien environment in which individual know-how and specialized equipment are more important than in any other area of special operations. He will learn about gauging his oxygen tanks and decompressing on his way to the surface. He will learn the effects that different sea temperatures have on his body – how he can swim on for ever in a warm sea, but tires more quickly as the water gets colder. He will learn to work with ocean currents to help him swim or paddle a small boat over long distances. He learns that in underwater demolitions the pressure of lower depths multiplies the force of an explosion.

A newly initiated diver with an SAS water troop might be taken down into a cove beneath the Indian Ocean to see where the sharks sleep. He will discover them bunked up on the rocks, not too differently from sailors in a cramped submarine, and find that the ocean predators can sleep soundly for days. He will be amazed that a school of sharks normally get frightened and scatter when a man swims fast into them. But if he has as much as a drop of blood oozing out of a cut or bruise their aggressive instinct is aroused by the scent to the point of overcoming any fear, and they attack.

Trained divers in the special forces are called 'Sharkmen'. Although 'Hell Week' conditions SEALs not to sleep, Marcinko could snore off a drinking binge for up to twenty-four hours and then be fully alert and in shape to lead his Team Six through their daily routine of four-hour shoots, six-mile swims, parachuting, room clearing and assault practice. He insisted that his men carry a weapon at all times. He himself once turned up at a Pentagon meeting with a .38 tucked under his sock.

A larger-than-life character, 'Marcinko thought he was God', according to a SEAL officer who succeeded him as commander of

Team Six. 'He had the balls, charisma and most incredible ability to step over whoever was necessary to get what he wanted', whether it was 'modified grooming standards' or MP-5s for his Sharkmen.

Team Six was constantly on the move – either free-falling in Arizona, climbing oil rigs in the Gulf of Mexico or exercising in the Tasmanian Sea with the Australian SAS. To get them used to living covertly Marcinko insisted that they drive in hire cars and stay outside military bases while on the road, climbing up the walls of their motels to get into their rooms at night for further practice.

The type of terrorist likely to hijack an ocean liner or take over an oil rig would have to be highly skilled and trained to the level of most special forces. The leader of a Latin American international terrorist organization, the Ejercito Revolucionario del Pueblo (ERP) describes the 'special operations' instruction he received at a training base in Cuba. 'The first three months of basic training consisted of constant practice, mainly with weapons from the capitalist bloc. American, French and Israeli arms were favoured as a way of covering up the command and logistics for our operations. There were courses in unarmed combat and demolitions with improvised explosives which could be made of ammonia, butane gas and various incendiary chemicals and industrial explosives like plastic, dynamite and TNT.

'Three months followed in a specialized phase. Rural and urban guerrilla tactics were studied. Sabotage operations were taught using actual models of those carried out by the Viet Cong and other communist insurgent movements. We were taught to use mortars, anti-tank weapons and anti-aircraft artillery. There was close instruction in conspiratorial methods to include secret communications and various techniques of surveillance and counter-surveillance. We studied FBI and other US Law Enforcement manuals to know the kind of behaviour and appearance that their agents would look for when trying to uncover a terrorist or a drug dealer in an air or sea port so as to know what to avoid. The training was highly individualized and it was not uncommon for operatives to be trained entirely on their own. Cells of never more than five men would train at times for a period of months to undertake a specific mission. The specialized courses would often include water tactics and scuba diving.'

To counter this kind of terrorist threat, elite CT special forces units had to be even better and more deadly. 'Safety is out the window,' Marcinko would tell his troops. 'We will train the same way that we will fight – balls to the wall. That means some of you will get hurt training. Some of you will die. That's inevitable.' Two members of SEAL Team Six were killed in accidents during the first few months of the unit's training.

Marcinko did not see terrorism as part of an ideological struggle between East and West but rather as a far deeper struggle that would outlast the Cold War – one 'between anarchy and order, culture against culture, sociopaths against society'. In effect, he wanted his men to learn to be and think like terrorists. He let his own beard grow down to his chest and his hair down to his shoulders, tying it in a long pony-tail. He insisted that all his men follow suit and get pierced ears so that they would look like unemployed delinquents or workmen in order to infiltrate inconspicuously into the communities where terrorists might thrive or the dockyards and industrial sites that could be their targets.

As Navy and Marine commando units were brought into counter-terrorism, it inevitably led to overlaps and duplication with Army units already tasked for the role, provoking rivalries between them. Among the many precedents which Marcinko broke was requiring Team Six to qualify in High Altitude High Opening (HAHO) and High Altitude Low Opening (HALO) parachute training, which had until then been largely the preserve of Army special forces. In Britain, the SAS rushed to upgrade their expertise in water operations as the much smaller SBS prepared for CT. They purchased a practice mini-submarine which could be used in a pool – something the SBS did not have. The SBS often resented the SAS swimming into their waters. 'Why train people for something in which there are already experts?' Ram Seagar would ask.

Rivalry between the SAS and the SBS even reached the point of Byzantine debates at the joint command level of British Special Forces; for example, if a ship docked in a harbour is the target of a terrorist attack whose responsibility does it become? The SAS would argue that if it's not at sea it becomes a building and is therefore theirs. As far as the SBS is concerned, a ship is still a ship.

When a terrorist bomb threat was received on board the *QEII* during the middle of a transatlantic voyage, it was an SAS man who was sent parachuting into the ocean with a Royal Navy bomb-disposal team to rendezvous with the ship. It was quicker to get them there by plane than by boat or helicopter and since the SBS lacked airborne training at the time, the SAS stole the show.

But Seagar has little doubt about who will get first crack if Rambo ever reaches the North Sea. When the Army for the Liberation of the Arctic Circle (ALAC) storms on to an oil platform, holds all the crew hostage and threatens to send thousands of barrels of oil spewing up in flames, the SBS will immediately dispatch its swim teams by submarine and helicopter to within striking range of the platform.

As negotiations are getting under way with ALAC, which is demanding the recognition of an independent Eskimo republic in Greenland and the payment of billions of dollars in reparations by oil companies for the environmental degradation of the Arctic, an SBS reaction force of about forty men is assembled in an airbase hangar or on board a naval vessel, depending on the oil rig's distance from the coast.

They study all the available intelligence on ALAC and whatever background information is available on its Iranian-trained leader code-named 'Quinn'. They may find out about a previous terrorist attack in which he proved a crack shot with an anti-tank RPG bazooka. They also study in detail the layout of the oil rig, using engineering blueprints which are instantly accessible via an SAS–SBS database of strategic targets.

At a point when negotiations seem to be breaking down and ALAC threatens to start killing hostages, the SBS assault force goes on 'alert status', boarding four Royal Navy Lynx helicopters equipped to carry twelve men each. For the next hour, the pilots keep the engines on. Rotor blades turn slowly as the tension mounts. The order suddenly comes to get airborne and the helicopters take off into the grey, stormy skies, hovering towards the oil rig. Then they circle around some three kilometres from the platform before being ordered back to their take-off point. They land but the assault force are ordered to remain inside the choppers, feeling beads of cold sweat trickling down the rubberized collars of their fire-retardant suits.

Inside the crisis control centre the atmosphere is sombre as the SBS commander receives confirmation from the Pentagon that the serial numbers of Stingers which ALAC claims to be armed with on the oil rig match those of a shipment of the anti-aircraft missiles which the CIA lost while supplying them to anti-Soviet Mujahidin guerrillas in Afghanistan some years earlier. The terrorists could easily shoot down the helicopters and destroy the SBS assault force. A water insertion now has to be ordered to neutralize the threat.

On board a submarine patrolling sixty feet under the sea some four kilometres from the oil rig, twelve swimmers are closely examining engineering plans of the platform's understructure, through which they will have climb to from below the icy waters. The latest weather report indicates that the wind-chill factor is likely to keep temperatures on the surface just below freezing point. They are already in their rubber dry suits, which are designed to provide heat insulation by means of a lining of cellular air pockets. These pockets have the effect of a blanket of dry air, insulating the swimmers' bodies from the Arctic waters, which would normally freeze a man in less than an hour.

When the alert signal comes on, they immediately zip up their dry suits securely, fit on underwater goggles, strap their aqualung rebreathers around their chests, carefully laying them over shoulder holsters with stainless-steel revolvers. They slip on their carrier vests holding flashbangs and extra ammunition clips, sling MP-5s around their necks and make their way to the submarine's escape chamber, carrying 30lb backpacks holding explosives, waterproof radios and climbing equipment. Four at a time, they pass through the watertight door into the cramped little compartment with water trickling in and the beam of a pressure light illuminating all the tubes and pumping mechanisms. They adjust the rubber fins around their feet and a minute later water rushes in, submerging them totally before the escape hatch opens and they float out through a cloud of air bubbles into the freezing darkness of the North Sea.

Holding on to the outer casings of the submarine, they breathe from mouthpieces connected to a lashed-on oxygen tube, riding with the vessel underwater for the next few kilometres to their insertion point. Communicating with the submarine commander through a transmitter receiver attached to their backpacks – over a Special waveband which automatically corrects the Donald Duck speech caused by the helium–oxygen mix they are breathing – the swimmers are informed that they are a few hundred yards from the target. The submarine rises to a depth of thirty feet, where the swimmers can start using their rebreathers. They switch mouthpieces, let go of the casings and use their combat knives to slash the ropes of two Gemini inflatable boats with outboard engines attached to the exterior of the submarine.

After pulling their rip cords to inflate their life vests, the men slowly climb through the water with the deflated Gemini boats, stopping for up to a minute every few metres to decompress – clear their bloodstream of air bubbles which will otherwise burst and can cause blood clots, ruptured veins or exploding eardrums. The divers look at a chronometer measuring their depth against the amount of time spent below water and calculate their decompression stops accordingly.

Surfacing on to ten-foot swells, the Sharkmen pull rip cords to release built-in air valves that automatically inflate the Geminis, and pile on top with all their equipment. It is night and the swimmers will have to navigate by compass or using their Global Positioning Systems through the waves which can easily block their view of the oil rig until they are right next to the skeletal concrete-and-steel structure, some eighty feet high. They slip into the water from the Geminis, towing the boats loaded with equipment and swimming over breakers which can easily hurl an entire team way past its mark. Reaching one of the huge hull columns, a swimmer immediately lashes a rope around it, tying it to his leg so that he won't get swept away. Now begins the seemingly impossible task of climbing on to the ice-glazed metal surface of the platform.

Noise is one thing that is not a problem amid the crashing sea and howling thirty to forty-knot winds. The two best climbers claw their way up the slippery ice covering each column to the steel braces supporting the platform, to hook on caving ladders. One man falls back down into the water and starts again, finally reaching the lower steel brace and wedging in the titanium hook. The ladder rolls down for the other eight men to begin their forty-foot climb.

They have a route mapped out, through the understructure's bracing, to an emergency ladder which goes vertically up the side and on to the back of the platform near the helipad. They are puffing and wheezing from the strain of their frozen ordeal by the time they have climbed to a point a few feet from the railings, when the lead Sharkman spots through his NVGs the ghostly heat source of a human appearing close to the edge of the platform. Without hesitating, he draws his stainless-steel magnum and holding on to the ladder with one hand, fires at the head.

With the extra rush of adrenalin which the sudden danger has now pumped through their bodies, they complete their climb in the next few seconds and pull themselves over the railings and on to the platform. Unslinging their MP-5s, half the team rush to take control of the helipad, covering the rear of the platform from concealed positions as the radio man signals the helicopters circling some three kilometres away to land the main assault force. The other six men move at lightning speed through the platform's superstructure to take the control room serving as the terrorist command centre.

As with skyjacking, maritime CT is so complex and dangerous – as well as being most likely to take place on international waters – that units from different countries may go into an 'international team mode', joining forces for an intervention. SEALs, SBS, Australian SAS, GIGN, GSG9 and other special forces which train for marine CT rehearse together regularly to standardize techniques. Combining their forces throughout all facets of an operation, even room-clearing 'shooter pairs' may be composed of men from two different national units. Launching a counter-assault on a hijacked cruise ship may involve rescuing well over a thousand hostages while clearing hundreds of small cabins, large-sized staterooms and highly explosive engine rooms. Cunard, the Grace Line and other shipping companies regularly allow CT units to practise on their ships.

On their final approach to a target, the first two helicopters carrying the assault force skim the waves at a low altitude below or level with the main deck. The chopper rises quickly to some sixty feet over the stern of the ship, letting the SEAL team drop by rope on to the deck. 'Fast roping', in effect a controlled fall by twelve men in seven seconds, can go drastically wrong in the small landing zone at the rear of a vessel tossing about in a rough sea. It requires coordinating split-second moves to avoid brain concussions and even more serious accidents. 'You have to learn to tug the rope just a bit as you hit the bottom,' says one SEAL Team Six commander, 'but not so strongly that it blocks others coming down above you.' He remembers once hitting a deck a fraction of a second before his radio man loaded down with 50lb of equipment practically piled on top of him. 'I could feel his body and backpack scraping the side of mine. If I had moved one decimal of a second slower, I would have ended in hospital.'

The next moment two other helicopters carrying a backup team and snipers are hovering dangerously close to each other above the ship, their rotor blades barely missing its funnel and masts. Most of the assault force disperse in pairs throughout the interior of the cruise liner while a full boat crew of six men immediately head up to the staterooms. Unlike in a building on land, where close-in surveillance and electronic methods may be used to determine the location of hostages, this is virtually impossible to accomplish on a ship at sea.

The SEAL team has only dated information which came in during the initial stages of the ship's seizure when passengers were concentrated in the dining area.

Reaching the upper floors after climbing several flights of narrow stairs, three men burst into the open dining room while the team commander leads the other two further down into the adjoining ballroom. They hear gunfire and screaming coming from inside, the shotgunner blasts open the lock on the double doors and the SEALs rush inside in tiger-stripe jackets over black wetsuits, their faces covered in Balaclavas.

'Get down!' screams the commander, aiming his MP-5 as nearly a hundred panic-stricken hostages, most of them still wearing their evening clothes, start running from the centre of the room in all directions, jumping or stumbling over piles of upturned furniture. There is gunfire coming from terrorists hiding in the middle of the crowd. 'Whoever doesn't comply with the order gets zapped,' explains the SEAL team leader as he and the other SEALs moving around the sides fire their sub-machine-guns into the centre of the ball room. The space is too large for stun grenades to have much effect.

SEAL Team Six was called to the 1984 hijacking of the *Achile Lauro* cruise liner, on which 427 passengers were held hostage by the Palestine Liberation Front. They were immediately flown out to a NATO airbase in Italy and from there helicoptered on board the assault ship USS *Iwo Jima*, which began shadowing the *Achile Lauro* in the Mediterranean. It had been decided from the start to launch an operation to retake the 'Love Boat' if negotiations failed for the immediate release of all the passengers. The US Army's Delta Force stood by to assist the SEALs. The Italian Navy's Commando Ragruppamento Subacqueri Incorsori (COMSUBIN) would also be supporting Team Six, although Italy's naval special forces unit had undergone little CT training.

SEAL officers closely monitored the negotiations over the next two days while studying local sea currents and all the available engineering plans of the *Achile Lauro*. Despite efforts by the Egyptian government to act as intermediary, negotiations seemed to be hopelessly stalled by the third day. Influencing the negotiations to their advantage, the SEALs had managed to get the ship manoeuvred so that it was positioned at the correct angle to the moon to facilitate a night-time helicopter landing on its deck. Having completed all essential preparations, Team Six boarded their helicopters on the flight deck of the *Iwo Jima* and sat inside them with the engines running and rotor blades whirring as swimmer teams sped off in their Zodiac boats towards the cruise liner. At the last moment, however, news was received that the terrorists had abandoned the liner, transferring to a small boat to reach Egypt, where the government had agreed to grant them temporary asylum.

The PLF had taken advantage of the difficulties involved in monitoring events on board a hijacked ship in high seas. Without the SEALs or the negotiators finding out, they had shot one of the hostages, Leon Klinghoffer, a sixty-nine-year-old wheelchair-ridden paraplegic from New York, on the second day of the hijack and dumped his body overboard. When the American Ambassador to Egypt, Nicholas Zeliotis, found out about the killing after boarding the *Achile Lauro* the minute it docked at the Egyptian port of Alexandria, he sent an urgent cable to Washington: 'You tell the [Egyptian] Foreign Ministry that we demand they prosecute the sons of bitches.'

But what could have been the biggest counter-terrorist coup since Princes Gate four years earlier turned into an Italian farce. By the time the State Department had completed the necessary paperwork to make an official request to Cairo for the arrest of Abu Abbas and the other terrorists, the PLF were on an Egyptair flight to Algeria, escorted by Egyptian security men. From his White House basement office, the National Security Council Officer responsible for counter-terrorism, Lieutenant Colonel Oliver North, immediately got President Ronald Reagan's approval to use US Navy jet fighters to intercept the airliner and divert it to a NATO base where the terrorists could be arrested.

Streaking over the Mediterranean, two F-14 Tomcats from the aircraft carrier USS *Saratoga* diverted the Egyptian Boeing to the Italian Air Force base in Signorella, southern Italy, where SEAL Team Six stood by to make the arrests. They were under orders to shoot the Egyptian commandos escorting Abu Abbas if they put up any resistance. But as the armed SEALs surrounded the airliner taxiing to a stop on the runway, Italian Carabinieri of the counter-terrorist Grupo Speciale d'Intervenza came racing on to the runway, jumping from their cars and vans to form a concentric circle around Team Six. The Italian government was insisting that they had criminal jurisdiction over Signorella. There was a tense stand-off as the SEAL platoon leader on the runway radioed for permission to shoot on the Italians. It was denied and the Carabinieri were allowed to make the arrests.

A group of SEALs flying by helicopter followed the Carabinieri aircraft, which carried the PLF terrorists to Rome. After Abu Abbas and his gang were seen entering prison, the SEALs were finally stood down and withdrawn from Italy. But some days later, Abu Abbas was inexplicably released from jail and put on a flight for Libya.

When he took command of SEAL Team Six a few years after the *Achile Lauro* incident, Commander Ron Yaw demonstrated to a group of visiting Congressmen what might have happened on board the cruise ship's cabins if word of Klinghoffer's death had been received a day earlier. The VIPs were sat down as hostages in a small killing room beneath cardboard dummies painted to represent terrorists. The lights went out, smoke came up through the floor and a voice was heard over a loudspeaker: 'You are now prisoners of the Palestine Liberation Front.' The next second, Commander Yaw and two other SEALs burst through the door. 'Pfft, Pfft . . . Pfft, Pfft . . . Pfft, Pfft' was all that was heard during the next one and half seconds before the lights came back on. As Yaw lifted off his NVGs and holstered his silenced pistol, the Congressman sitting in front looked at the terrorist dummy and turned back to him, remarking, 'You missed him.'

'Just look a bit closer,' replied Yaw. Standing up from his chair, the Congressman got right next to the cardboard figure, then turned around looking amazed. The SEAL's double tap had diametrically pierced both eyeballs.

As Yaw says of the SEALs, 'We like to be in close combat – pistol combat, knife combat, the closer the better. The other guy represents Father Death.' Delta men refer to the SEALs somewhat derisively as 'real studs'. Team Six patronizingly call their Army cousins 'college boys' because of the emphasis which the unit has developed in academic qualifications, their widespread use of psychologists in both selection and training. But the two units' scores are very close when it comes to shooting competitions. If anything, the SEALs tend to do better in the twenty-five-metre pistol shoot while Delta's snipers generally win the long-distance rifle event, having greater proficiency in high-angle shooting.

Similarly, units of many different nationalities and employing a variety of tactics and training methods, converge once a year to play in the highly classified CT Olympic Games. They see who can rope up the wall of a building the fastest; who can move the fastest through a killing house, using manholes, ventilator shafts and underground passages, then clear a room of targets rotating around a female figure in the shortest possible time.

There are obstacle races over tree trunks, karate and judo competitions, swimming events in which a man ropes on to a diving board, from which he dives into a pool to recover an object at the bottom and then swims underwater to the other end. In one particularly awkward competition, a pair drive a Land Rover around an obstacle course. The driver is blindfolded and follows the directions of the other man. It's all timed, down to milliseconds. The GIGN, which once won third place, thinks that it is 'the wrong philosophy. It's purely physical without any emphasis on the calculated mental approach required in real CT work.' The SEALs and GSG9 participate in the Games every year, usually getting first prize or finishing in the top three. Delta is often represented, having once carried off the top award. But 22 SAS refuse to attend, considering the whole thing to be quite beneath their dignity.

'Special forces are made of men who like to live life on the edge. Whose sense of challenge is not fulfilled without the presence of Father Death,' says Ron Yaw. 'It's facing a very dangerous situation and conquering it. It's overcoming the maximum adversity which gives you the ultimate satisfaction. We want to be in a situation under maximum pressure, maximum intensity, and maximum danger. When it's shared with others it provides a bond which is stronger than any tie that can exist between men.'

Bo Gritz, who once swam across the Mekong River in search of missing American POWs in Laos, writes that 'Combat contains a measure of exhilaration seldom equalled in any competitive sport, the stakes are high and the losers die.'

9

THE INSERTION

A few years after his military career exploded, together with a C-130 and Marine helicopters in Iran, retired US Army Colonel Charles Beckwith sat as a reluctant celebrity at a Hollywood studio. Big movie producers had flown him up to Los Angeles to discuss making a movie based on his recently published autobiography, *Delta Force*. A multimillion-dollar budget was being proposed and big-name stars were lined up. 'The only problem is the ending,' said the executive producer matter-of-factly, as he leaned over the conference table puffing on his cigar. 'We'd like to change it so that you get to Tehran and free the hostages.'

Despite his acquired worldliness and the proximity he had enjoyed to the pinnacles of American power, deep down inside Charlie Beckwith had never stopped being the hard, simple soldier from Georgia who always kept shining his boots. The values of duty, honour and country, however trite they may have appeared in the cynical and materialistic eighties, continued to be the north star of his life.

Many accused him of loud vanity but what Beckwith had just heard came as a slap in the face to his sense of integrity and to the memory of those who died. 'I shouldn't really be here,' he muttered, half to himself, as he rose from the comfortable leather chair and walked out of the meeting, leaving the Hollywood men to remake Rambo's story on their own.

'I could have shot the goddamn helicopter pilot,' said Beckwith later, sitting in the living room of his modest suburban home in Austin, Texas, still bitter some thirteen years after the horrors and frustrations of that terrible night of Delta Force's abortive first mission. He recalls instinctively reaching for his .45 automatic pistol when, after arriving fifty-seven minutes late for their RV, the commander of the Marine helicopter squadron 'turned to look at me as I came into the cockpit and advised aborting the hostage rescue mission'. Two of the eight helicopters had got lost on the way and returned to the aircraft carrier *Nimitz* in the Persian Gulf. The nerves of the untrained crews, who had never flown deep into enemy territory amid hundred-foot sandstorms, were shattered.

Major Bucky Burruws immediately started loading Delta's ninety operators and additional support personnel, including Iranian drivers and translators, communications specialists and other technicians on to the six helicopters which had flown in. He used scales to weigh every one of the 120 men as they got on board. Knowing how much each RH-53 Sea Stallion could take, Burruws calculated that the six were filled to capacity. One of the choppers had exactly one pound less than its maximum weight. The lightest one was only seven pounds underweight.

Much else seemed to be going wrong from the moment that the C-130s carrying Delta Force and fuel for the helicopters had put down on the improvised landing strip which the CIA had picked as the staging point for the rescue operation in Iran. The first thing which struck Delta's Japanese-American intelligence officer, Wade Ishimoto, when he got off the aircraft, was the volume of midnight traffic on the highway within clear view of the landing site. When he went to gather up a squad of Rangers who had been brought along to provide perimeter security, Ishimoto found them all vomiting from airsickness. Only two were in any condition to drive up with him on their motorcycles to intercept a bus which they forced off the road by firing M203 grenades. Then Ishimoto noticed another set of headlights coming up the highway and jumping on to the back of a motorcycle behind a Ranger, ordered him to intercept the vehicle.

Drawing close, they realized that it was a huge truck. 'Get off the fucking road!' shouted Ishimoto.

'Which way, sir?' asked the Ranger.

'To the left! South!'

After ordering the Ranger to cover him with his LAW, Ishimoto got back on to the road and signalled the truck to stop. When it didn't, he fired his M-16 into the radiator. 'I had a mental flashback to my time in Vietnam when four North Vietnamese Army trucks rolled towards me as I fired away with an M-60 machine-gun. The Iranian truck now did the same thing.'

'Rubio!' Ishimoto called out the Ranger's name. 'Cock your LAW.'

'Yes, sir.'

'Are you ready?'

'Yes, sir.'

'Fire!' Whoosh. Both men hit the ground, letting out a simultaneous shriek of 'Holy shit!' as a fireball of flame and smoke exploded 100 metres into the night sky. They had just hit a fuel truck. A van following the truck screeched into a U-turn, picked up the driver, who was fleeing the blaze, and sped away. Rubio kick-started his motorcycle after two or three tries and chased the van, but turned back after a mile, unable to catch up.

Ishimoto broke open a bag of raisins and passed them out among the other Rangers, who had just caught up with him and were complaining that they were hungry after throwing up their lunch. This is great, he thought. We have food, a nice fire, a lot of Iranians could be dropping in soon and the helicopters which are supposed to take us to Tehran aren't even here yet. This is really our day.

Richard Marcinko was listening to all of what was going on live from the special classified intelligence facility on the Pentagon's second floor where the radio traffic in Desert One was being monitored via satellite. The Joint Chiefs of Staff, CIA people and a few handpicked special operatives, including the SEAL Commander, were part of the exclusive audience. 'I realized at that moment what the tension must have been like for the mission-control people in Houston as the moon lander left Apollo 11 and started its drop on the lunar surface. There was nothing they could have done if anything went wrong. There was nothing we could do if something happened to Delta either.' Marcinko defines the three stages of a special operation as (1) SNAFU (Situation Normal All Fucked Up); (2) TARFU (Things Are Really Fucked Up); (3) FUBAR (Fucked Up Beyond Any Repair).

The Iran hostage rescue operation had progressed to TARFU in a very short space of time and Delta would be lucky to limit it to that, in overloaded helicopters with shaky pilots who, because of their hour's delay in getting to the RV, would not be reaching Tehran before daybreak. But FUBAR came much sooner than expected. Burruws had just loaded the men on to the Sea Stallions when the Helicopter Squadron Commander reported to Beckwith that one of the choppers had developed hydraulic problems and couldn't take off. He again recommended aborting the mission. 'I got so mad that I couldn't see straight,' says Beckwith, who went to try to find the Air Force commander in charge of the C-130s, to get his opinion.

One of the USAF pilots landing a fuel-laden C-130 Talon at Desert One, Captain J.V.O. Weaver, had since Vietnam served in the 20th Special Operations Squadron (SOS), where he had operated an EC-47 electronic warfare plane. While the SOS planned their part in the hostage rescue operation 'in isolation' at their base at Hulburt Airfield on the northern Florida panhandle, Weaver remembers discovering many deficiencies. 'The communications were poor, we didn't really have the equipment we needed and command and control was confused.' Radio communications had been configured so that none of the three main components of the operation – Delta Force, the Marine helicopter squadron and the SOS – could talk to each other by radio when out of earshot without going through the mission 'coordinator', General Vaught, sitting back in Cairo. 'It was a perfect example of top-heavy planning.'

Through his windscreen above the luminescent dials of the C-130's otherwise dark cockpit, Weaver could see Beckwith and his Air Force commander, Jim Kyle, gesticulating wildly, arguing about who was in charge. All the time, aircraft engines were running hot, blowing sand and dust everywhere, blurring everyone's vision. Iranian prisoners who had come off the bus were sitting around guarded by the Rangers. The burning fuel truck on the highway lit up the night. 'It was a mess.'

The commanders, including General Vaught in Cairo, agreed that it had to be Beckwith's decision whether to go on to Tehran with only five helicopters, which inevitably meant leaving some people behind. If he left the Iranian drivers and Farsi translators, how would Delta get through the road checks they expected to encounter on their final approach to the US Embassy in vehicles? If he eliminated twenty men from the assault force, how would he have enough to clear at least four large buildings in a twenty-acre compound, secure a football stadium a block away and get the fifty-four hostages there for the helicopter evacuation?

But perhaps most importantly, what would happen if any more helicopters broke down? Beckwith realized that the pilots weren't really with him. They were edgy, their attitude negative. 'I needed real mavericks,' he told me one year before he died of coronary heart failure at the age of sixty-two. 'Guys who would say I'll go through with it even if I have to carry the goddamn helicopters to Tehran. I thought back to the Vietnamese helo pilot I had back in Nam for whom no mission was too dangerous. We just didn't have that.'

The main purpose for the complicated RV at Desert One was to refuel the Marine helicopters so that they could continue with their mission after their long flight into Desert One from the *Nimitz*. The only way to do this was with dangerously flammable rubber fuel blevits, resembling giant water-beds, which were offloaded from the SOS C-130 Talons and connected to the helicopter fuel tanks. This awkward process now led to a disastrous ending.

As the pilot of an RH-53 Sea Stallion gunned his engine after topping his tanks from a blevit very close to one of the C-130s, his rotor blades struck the plane's fuselage. There were lots of sparks, which set the fuel on fire, and then a huge explosion as the C-130 and the helicopter were incinerated. Ammunition and rockets carried on board the aircraft shot up into the air like fireworks and it was only the fast reactions learned in the 'killing house' which saved twenty members of Delta inside the burning plane, who managed to race out before being consumed by the inferno. The three crewmen on the helicopter, and five on the C-130, were burned to death.

There was no time for an orderly withdrawal. Delta men, Rangers, remaining helicopter crews and others ran into the surviving C-130s, which took off ASAP. Wade Ishimoto, who was still watching the highway, almost got left behind, being picked up by Jim Kyle in a jeep as the last aircraft was preparing for take-off. Three Sea Stallions were abandoned intact for the Iranian Air Force and Beckwith could not even fulfil President Carter's last personal instruction at their White House meeting of three nights earlier – 'Bring back any dead American bodies with you.' The charred remains of the eight airmen would be displayed before the television cameras by defiant mobs making warbling noises in front of the US Embassy in Tehran.

Beckwith burst into tears on the C-130 flying out of Iran, confessing, 'I've embarrassed this great country of ours . . .' His two executive officers, Burruws and Ishimoto, lay in sombre mood on mattresses on the steel floor of the fuselage, seriously discussing retirement. When he got back to Fort Bragg, Delta's commander was immediately ordered to Washington, where President Carter, who was seeking re-election, insisted that he give a press conference explaining his decision to abort the mission. 'Would it be conceivable that Margaret Thatcher would have asked Michael Rose to give a public press conference if some SAS operation had gone drastically wrong?' asks a former member of Delta.

Like a frightened child, Beckwith ran down the Pentagon's endless green corridors to find some general, some friend, who could stop that press conference. There was nothing anybody could do. The President wanted it and Beckwith had to enter the glare of the television lights and face the machine-gun fire of reporters' questions. 'There were some people who felt that you and I were at an impasse about conducting this mission,' explained Carter when Beckwith was taken over to the White House afterwards. 'I did not want to put you in front of the press but I really had no alternative.' At the very moment in which Beckwith could have been the biggest American war hero since Patton, he was, instead, the fall guy of a disgraced administration.

If the Carter administration had created Delta Force as a kind of insurance policy against terrorism they had neglected to read the fine print. Carter himself never took much of a personal interest in the unit, only going to visit it after the failed raid in Iran. However well-honed Delta's specialized skills may have been, other factors essential to direct-action counter-terrorist missions had been ignored and the type of understanding required at high levels to address these insufficiencies was just not there.

Since the regime of the Ayatollah Khomeini had taken power in Iran following the overthrow of the Shah in 1978 there had been plenty of warnings about what was to come. The US Embassy in Tehran had been the target of repeated attacks by uncontrolled mobs of radical Islamic militants and the State Department seemed to be ignoring this while actually lobbying American companies to reinvest in Iran. Almost a year before the hostage crisis a seizure of the Embassy had taken place, in February 1979, when demonstrators scaled the walls and occupied the compound for several hours. 'We knew intuitively that it would only be a matter of time before it happened again,' says Wade Ishimoto, 'and Delta immediately asked for permission to travel to Tehran to do a detailed survey of the US Embassy and its surroundings.'

Delta was working at the time with the Defense Advanced Project Agency on a high-tech interactive TV system called 'surrogate travel' which consisted of information in a database coupled with graphic displays on an optical disk. One could quickly find a location on a sketch map, mark it with a cursor and the corresponding pictures would show up on a computer screen. Using a joystick, one could then simulate walking down a hallway or into a room, then looking up at the ceiling, or follow other routes. But at the heart of the system was the ability to graphically record details of facilities like Embassies through spot surveys.

Permission for Delta to send a survey team to Tehran was denied by the Army and despite some appeals no political pressure was exerted to counter the order. 'No reason was given,' says Ishimoto, and 'the inability to survey the Embassy meant that there was no documentation on the Embassy buildings and surrounding area when it fell in November 1979, greatly complicating the task of planning a rescue operation.'

Similarly, the helicopter team assigned to Delta was chosen as a result of decisions taken in the Pentagon that had more to do with inter-service rivalries than with the mission's main requirement: to get in and out of a landlocked enemy capital in a country surrounded by other states hostile to the USA – the Soviet Union, Soviet-occupied Afghani-stan and Iraq. Since the helicopters had to take off from an aircraft carrier, the Navy insisted that they had to be Navy helicopters because Army ones were not equipped with folding tails and rotors, which would have allowed them to be stored below deck. Beckwith requested pilots with special operations experience who could be found in the Air Force SOS and some Army aviation units. But the Navy insisted that they had to be Navy pilots. The only previous experience of the flight crews initially assigned to Delta was in minesweeping operations and in their first joint rehearsal they proved so entirely inadequate that Delta had to insist that they be replaced.

But instead of agreeing to give Beckwith the pilots he wanted, it was now a chance for the Marines to get a slice of the action in a special operation which would make history. A Marine helicopter team 'which was picked off the shelf', according to Beckwith, with no special operations training or experience of any kind, was finally chosen by the Joint Chiefs of Staff. Beckwith says that he eventually came to feel 'comfortable' with the assigned helicopter crews. Burruws, however, felt that 'the pilots, quite frankly, never really expected to get to the execution phase of the mission. The plan was really foreign to their minds and if you don't believe that you are really going to do something you half-ass it.'

Christian Prouteau of the GIGN, which supplied some special tear-gas to Delta on their way to Iran, believes that 'Charlie never really had command and control of the operation. He should have said at the very beginning, give me all the decision-making power, don't use me as a weapon only.'

'I guess I ain't all that bright, you see,' muses Beckwith, recalling all the other pressures he was having to deal with. One example was 'the State Department trying to tell me not to kill the Iranians guarding the hostages in the Embassy, to only shoot to disable them. The debate went on right up until the White House briefing in front of President Carter when Undersecretary of State Warren Christopher again asked me what we intended to do about the guards and I flatly replied, we're going to put two bullets right between their eyes, sir. Carter didn't object and the matter was finally settled.'

There has been much disagreement about the hostage rescue plan devised by Beckwith. It has been suggested that it was overcomplicated, that Delta Force should have gone by parachute or driven into Iran overland from Turkey (which, although a NATO ally on Iran's north-west frontier, refused to support an American rescue effort when it was approached on the matter, fearing Iranian retaliation).

When Michael Rose, flush with his victory at Princes Gate, came to Washington to pay his respects to Beckwith, he unrolled his plan showing how the SAS would have done it. Rose was not entirely uncritical of the American. He believed that 'Delta's ethos was rather too rigid and muscular' and its commander tended to be 'inflexible', filling the unit with too many muscle-bound Rangers. 'Too much muscle tends to make for the wrong power-weight ratio and generates the wrong psychology,' according to Rose. Even so, the plan which the SAS had worked out in its spare time was virtually the same as Delta's. After the initial RV at Desert One, the helicopters would move the assault force to a hideout area outside Tehran where they would rest for the next day. That evening, they would be met by transport vehicles arranged through undercover operatives already in place and be driven to the Embassy.

Any sentries outside the compound would be taken out with silenced pistols as one team approached from the front, blowing out the padlocked gates, and another came in from the rear, slicing the back wall in half with demolitions charges. The main force would clear the four buildings in which the hostages were expected to be while a five-man team knocked out the Embassy's electrical generator. They would move the hostages to the nearby football stadium and from there they would all be extracted by helicopters which would fly in from the hide-site to take them to a final evacuation point. This was the only stage of the plan on which Delta and SAS differed. Beckwith had chosen the unused Mazariyeh airstrip, about half an hour from Tehran, which would be held by a Ranger company so that two giant C-141 Starlifters could be landed to fly out the hostages and Delta Force. Rose would have chosen an airstrip further away from the capital. 'But without adequate helicopter support,' he agrees, 'there is no way that the operation could have succeeded.'

Delta had rehearsed their planned break-in and room clearance of the US Embassy in Tehran no fewer than seventy-nine times. Bucky Burruws says he got 'to the point where we could do it blindfolded', adding, 'There was no doubt in my mind that once we got to the Embassy we would break in and get the hostages to the evacuation point in the soccer stadium. I was going to be up in the press box of the football stadium to direct in air support. But after that none of us could be too sure. Successful completion depended largely on the helicopters, which failed us at the beginning.'

'My recommendation is to put together an organization which contains everything that it will ever need,' Beckwith told a hearing of the Senate Armed Services Committee chaired by Senator Sam Nunn from his native Georgia. 'An organization which would include Delta, the Rangers, Navy SEALs, Air Force pilots, its own staff and its own support people, its own aircraft and helicopters. Otherwise we are not serious about combating terrorism.'

The incoming Reagan administration gave its full backing to legislation introduced by Senator Nunn to create a new Joint Special Operations Command (JSOC) which would coordinate the training, planning and requirements of special forces units in all three branches of the armed services. Based in Fort Bragg under an Army general, JSOC would report directly to the Joint Chiefs of Staff. It would have its own funding, independently administered through the Pentagon's new office of Undersecretary of Defense for Special Operations and Low Intensity Conflict, to prevent the different services from arbitrarily cutting back on expenditure on special forces, as was done after Vietnam. A major portion of the expanded budget would go towards upgrading the Air Force Special Operations Squadron, integrating it more closely with Army and Navy components of JSOC so that the disaster at Desert One would not be repeated.

'For the Pentagon, this was a radical new departure,' says Richard Marcinko. It was considered 'heresy' by important elements in the top brass. J.V.O. Weaver, who as an SOS C-130 pilot had personally experienced the disaster at Desert One, saw that 'there was tremendous learning to be done between the different services. The Army had to learn that the Air Force was more than a bus service. That getting around ground radar and SAM missile sites while going into enemy territory was a very fine art.' While he was writing a paper outlining the proposal to interface the different special forces units, he was called in by the Vice Chief of the USAF, General Carns. The large, intimidating man sat back on his leather chair behind his desk, with three stars shining over each shoulder patch on his neatly pressed sky-blue shirt. 'Hey, J.V.O., I've heard a lot about you, have a seat . . . We understand that you are writing an important piece of doctrine and this is the way we want it read: Air Force assets should only be available to special operations forces on a mission-by-mission basis. No Army guy should ever have control of an Air Force asset. Is that clear?'

Weaver disagreed, expressing his heartfelt conviction that command and control of special forces had to be restructured since that was the kind of war which the USA was most likely to be fighting in future. Cairns just exploded. 'Get the fuck out of here!'

One of the problems with getting special forces taken seriously at the Pentagon had always been that they were basically about 'small bucks'. In the world of multi-billion-dollar defence budgets, diverting a few million for the 'snake-eaters' was often looked upon as, at best, a nuisance. Generals and Pentagon *apparatchiks* had little motivation to release and administer funds which would not lead them into the revolving door towards plush jobs and consultancies with the major defence contractors selling tanks, jet fighters and battleships. It was a case of the banker who makes million-dollar investments not wishing to be bothered with thousand-dollar loans. 'They're never interested in simple solutions,' says Bo Gritz. 'What they want is one-hundred-million-dollar ones.'

The SEALs had been started with less than two million dollars; Delta Force with less than four million. Despite inflation, the entire annual special forces budget had barely topped the hundred-million-dollar mark set by President Kennedy back in 1961. But creating a special operations air force was going to take real money. Shortly before the Iran crisis broke the USAF SOS was based in a sleepy little airfield in northern Florida which had not changed much since it was a naval air training station in World War II. The airfleet at Hulburt Airfield consisted of a few C-130 transports and gunships, some Vietnam vintage UH1 helicopters with no more than two hours' flying capability and a group of eccentric pilots who enjoyed making blind landings at night. 'We were like the bastard children,' says one of them. But the series of legislative initiatives during the eighties which brought special forces budgets into the billion-dollar bracket changed all that. By 1993, over 300 million dollars, almost a third of the annual special forces appropriation, was going towards the continued upgrading of the Air Force component.

'Our main priority,' says Dr Alberto Cole of the office for Special Operations and Low Intensity Conflict, 'is to make sure that we have the capability to insert into any location, anywhere in the world and get out again in one night.' The Lockheed Corporation, Boeing, United Technologies, Rockwell International, IBM, Texas Instruments, had all now been brought in to develop the new special operations airfleet: helicopters, gunships, transports and fuel tankers, all specifically designed to fly under any weather conditions, at low altitudes and deep in hostile airspace.

A squadron of C-130 Talon Tankers now provide in-flight refuelling for helicopters, a revolutionary new concept, despite its simplicity, which would have prevented the tragedy at Desert One had it been available at the time. Until recently, in-flight refuelling was the exclusive preserve of high-performance jet fighters and supersonic long-range bombers, conducted at high altitudes and 500mph velocities from giant KC-135 Tankers. 'The main thing we have to do is to train to think helo,' says one of the new generation of SOS Tanker drivers. 'We have to know what their needs are even before they do.' A four-engined, propeller-driven C-130 has to slow down to its minimum safe speed of 120mph at altitudes which may go as low as 500 feet for a helicopter to inject its fuel probe into the eighty-foot hose extending down from the wing tank. The manoeuvre must be accomplished at night while flying on instruments and seeing through NVGs.

'Over enemy territory we have to find a "blind zone" which is relatively clear of enemy radar and anti-aircraft sites to conduct the refuelling operation.' To manoeuvre around enemy radar screens, the Talon Tanker is equipped with an Electronic Counter Measure (ECM) suite operated by one of its crew which picks up enemy radars, identifying the different types, whether they are scanners or missile-guidance systems and varying the aircraft's flight path accordingly. The ECM simultaneously feeds deceptive electronic information to the enemy, giving false range, speed and altitude.

Sitting in the press box of the football stadium in Tehran, Bucky Burruws would be calling in fire from AC-130H 'Specter Ghosts' circling some 15,000 feet in the night sky, protecting Delta's helicopter evacuation of the hostages from an Iranian reaction force. A computerized system of interfacing sensors built into the special operations gunship can direct fire with pinpoint accuracy on several ground targets simultaneously. Along with the 105mm howitzer, twin 40mm Bofors cannon and a 25mm Gatling gun sticking out of one side of the fuselage, there is a hump which looks like a giant mole, containing the all-important sensors. The signals and images these pick up are fed to screens in the corresponding 'sensor shacks' inside the aircraft.

Sitting in his cubicle, the electronic sensor operator sees the distinct smudge of a tank's engine outlined by its electrical charge. The information is cross-checked with the infrared sensor detecting the tank's heat signature and that of soldiers around it. Low-light TV, which in the newly modified AC-130U Specters can operate in any weather, also projects an image of the target on to a screen in front of the Fire Control Officer. With all the information processed by computer, a target laser designator directs fire from any of the guns on to the tank and enemy personnel. Twenty new targets can be entered simultaneously into the fire-control system as they are called in by operatives on the ground during a fast-moving battle. The in-flight computer can also automatically match the information picked up by the different sensors with pre-selected targets already stored in its memory. If the ultimate 'sniper in the sky' really has to see what it it's doing, a mounted spotlight is switched on which would light up an entire football field.

The cockpit of the latest MH-53J Pave Low Helicopter is a virtual spaceship. Flying a night-time infiltration mission through a sandstorm, the pilot scans his forward-looking terrain following/terrain avoidance radar for 'real time' radar video imagery of the land features, contours and objects which are directly beneath or miles out in front. The screen continuously shows up-to-the-second digital computations of the helicopter's altitude and range down to inches. The pilot then checks his moving map display, which gives a horizontal pictorial vision of the helicopter's position and movement relative to a computerized flight map. To find out exactly where he is in relation to a programmed target or waypoints in a flight plan, the pilot turns to his Doppler navigation system, which is constantly computing the helicopter's velocity and movement and triangulating the information with satellite data to give the precise distance and time it will take to reach his destination.

When something appearing on the radar screen requires his special attention, the pilot switches on his forward-looking infrared detector, which gives him an intensified television image of the aircraft, building or vehicle, showing a more accurate outline of its heat signature. The special operations Pave Low is the world's most advanced helicopter, incorporating radar and navigational equipment used until now only in F-15 and F-16 fighters, F-111 bombers and a few of the other most advanced jet aircraft. The SOS fleet of combat helicopters includes a squadron of MH-60 Blackhawks so heavily armoured that while inserting a Delta team during the 1983 invasion of Grenada one managed to continue to fly after taking forty-seven hits from 23mm anti-aircraft guns.

'With the kind of air-insertion capability we have today,' says Bucky Burruws, 'our rescue operation in Iran would have been far simpler to execute.' Pave Lows and Blackhawks, which are now built with collapsible tail and blades so that they can be hangared below the deck of an aircraft carrier, would have refuelled off C-130 Talons in mid-flight as they entered Iran flying below radar cover. 'We could have put down the assault force right on to the Embassy grounds. Specters could have shot up the telephone poles and other obstacles against a helicopter landing. In half an hour we could have had the hostages out of the buildings and on to the waiting helicopters – especially since we received last-minute intelligence that they were all concentrated in one building guarded by only one Iranian at night. The whole operation could have been conducted with half the amount of people and in a fraction of the time which was factored in our original plan.'

'Today, the Iran hostage rescue operation would have succeeded,' agrees General Sam Wilson, who as director of the Pentagon's Defense Intelligence Agency (DIA) throughout the eighties, pushed for upgrading the Air Force special operations capability just as he had worked behind the scenes to help create Delta Force. Among the many new buildings currently going up at Hulburt Airfield is an entire wing to house a section of the DIA. 'I went over to spend a great deal of time at the Air Force Academy in Colorado,' says Wilson. 'It was not enough to just have the right machines, but we also had to make sure that we recruited the right kind of pilots for special missions. They need to have the same daredevil qualities as supersonic bomber pilots who would strike deep into the Soviet Union or regular fighter jockeys flying F-16s up into the stratosphere. But with the difference that for deep infiltration missions they must like to fly slow and low, crawling below enemy radar cover. It's a tough combination to find in a pilot.'

'Are you willing to spend the rest of your military career on the road flying with night vision goggles at altitudes of 100 feet or less?' Captain Tommy Trask remembers being asked during his interview for the SOS. His answer was yes. 'The idea of special operations had always excited me because it meant having a closer personal association with national policy,' he explains.

Like Army special operatives, SOS pilots go through punishing training in survival, land navigation and resistance to interrogation. 'I almost changed my mind during parts of the training, seriously wonder how the hell I got through it,' comments Trask. 'In the SOF [Special Operations Forces] community, we consider ourselves mauve-card men, purple suitors, and want to feel on equal terms with the Army and Navy guys,' says the blonde, moustached, kind-eyed helicopter pilot.

When special forces teams planned their incursions into Iraq during the Gulf War, they did so together with the helicopter crews. Pilots and co-pilots tasked to insert, extract and resupply Army special forces patrols would go into isolation with the land teams, so that they would all get closely acquainted with each other's perspectives on the various requirements and problems involved in the operation.

Tom Trask and his six-man Pave Low crew had been on twenty-four-hour standby by the morning of 21 January 1991 in the desert airfield of Ar'ar, just fifteen miles south of Saudi Arabia's border with Iràq. It was only four days since the first American-led air bombardment of Baghdad and almost a month before Allied air supremacy over the skies of Iraq would be declared. The main SOS mission at this early stage of the Gulf War was search-and-rescue of downed pilots. At 7.30 a.m. a call came through on Trask's satellite radio that a US Navy F-14 jet fighter from the aircraft carrier *Saratoga* had been shot down on its way back from a bombing raid on Baghdad by a Soviet-made SA-2 missile.

Being totally fogged in, the airfield was operating under IFR (Instrument Flight Rules). Although half a squadron of MH-60

Blackhawks were also based there, only the Pave Low, with its special radar and all-weather systems, could operate effectively in the near-zero visibility. Answering their call sign of Moccasin Zero Five, Trask and his men scrambled into their helicopter. After fitting on their flight helmets and plugging in their built-in radio headsets, they strapped themselves into their seats and ran the routine checks. Lieutenant Lew Caporicci sat in the cockpit as Trask's co-pilot, manually testing the controls as the captain read out their checklist. The radio man tested all the frequencies, aligning the satellite link-ups. The left and right window gunners loaded and cocked their belt-fed .50-cal machine-guns, the flight medic got as comfortable as possible on his canvas seat and the pararescue man took his position by the helicopter's back ramp.

Trask would have preferred to fly this kind of IFR mission at night, which is what the Pave Low is specifically designed to do. Although it is fully capable of manoeuvring through enemy radar screens in daylight, there was a much higher risk of visual detection. But in darkness the F-14 pilot would, in all probability, be another hostage of Saddam Hussein's.

It was 7.55 a.m. when the down-draught of the rotors lifted the helicopter off the tarmac in the vertical forward motion which gives the sensation of going up on a Ferris wheel. Using the pedal controls, Trask turned the helicopter by its tail-rotor action and headed ninety degrees due north. Cruising in the thick fog, visibility was limited to less then than 100 yards as he choppered the Pave Low into Iraq. He was maintaining a speed of 140 knots and an altitude of 100 feet, when his right scanner spotted an Iraqi lookout tower just as they were practically on top of it. 'I realized that we had now been seen. Within five minutes of entering Iraq all enemy radars were looking for us.'

About fifteen miles from the border, the fog lightened a little and Trask dropped down to an altitude of fifteen or twenty feet. He called in to one of the AWAC planes permanently on station some 30,000 feet above them, monitoring enemy electronic screens. 'This is Moccasin Zero Five heading to survivor position.' Almost immediately the vectoring from a female voice came on the line, 'Moccasin Zero Five, come 30 degrees, SA2 radar is up at you one o'clock at ten miles.' Trask manoeuvred the Pave Low slightly west to avoid the missile-guidance system. Then he heard, 'Moccasin Zero Five, snap south, unidentified aircraft closing in on you.'

'The guy in the lookout tower has obviously called his boss,' Trask thought out loud. The Iraqis had just launched a fighter after him and the AWAC controller was suggesting that Trask turn and run back to the Saudi border. But she was obviously used to dealing with fast jets. She didn't realize that helicopters don't 'snap' anywhere. Trask calculated that at his maximum speed of 150 knots, if he followed her advice he would end up getting shot down about two miles from the border. The only thing to do was take evasive action. A rough semicircle of green smudges on his terrain-following radar indicated a shallow wadi up ahead, which he flew into, hovering at an altitude of twelve feet, just above the depression, so that the slight land elevation could protect him from the jet fighter's radar. 'Where's the F-15 CAP,' Trask called in to the AWAC, requesting assistance from USAF fighters.

Having kept the helicopter stable for nearly five minutes by constantly making the most minuscule corrections of the cyclic, collective and pedal controls, Trask suddenly saw the French-made Iraqi Mirage F1 shoot right over him. It flew straight past without seeing the helicopter and half a minute later he recognized two F-15s streaking out of the sky in pursuit of the Mirage. His manoeuvre had worked. He pulled the collective control to gain slight altitude to twenty feet as he continued flying beneath the fog, pushing north. The Pave Low's inertial navigation system indicated that he was now thirty miles inside Iraq.

Trask knew that only some radar systems can pick up an aircraft below 100 feet if it gets right on top of them. He was more worried about radio intercepts. Usually radio silence was maintained during deep infiltration missions. But in a search-and-rescue operation without a pre-programmed flight plan, radio contact with the AWAC had to be maintained at all times. He got a call again warning that two enemy helicopters had just taken off in his direction. 'It seems like everyone in Iraq knows we are coming,' he calmly told his crew over their headsets. 'My gunners were looking out at the sky, just itching to have a go at the lightly armoured Soviet-made MI-8s. But the Iraqi helos never found us.'

The Pave Low flew on for another 100 miles. The fog melted as the helo approached the map coordinates for the shot-down F-14. 'Slate 46, this is Moccasin Zero Five. Identify your position. Over.' The Pave Low's radio man, the AWAC and F-15 fighters providing cover, repeatedly tried raising radio contact with the downed pilot by his call sign but there was no reply. Slate 46 hadn't been heard from since his original mayday signal two hours ago. They moved further north and Trask began thinking that the survivor had been captured. The Pave Low circled around a wide area of desert for about twenty minutes, finding no sign of him.

'It was time to head back. The helicopter had just enough fuel to make it back to Saudi Arabia and with every Iraqi radar looking for us it was not possible to gain sufficient altitude to get mid-air refuelling from an HC-130.' On the way south, the AWAC directed the Pave Low towards the Iraqi airfield of Mudaysis. Intense anti-aircraft fire detected there could have shot down the Navy F-14 and its survivor might be in its vicinity. Dropping down to ten feet, below any anti-aircraft guidance system, Trask hugged the contours of the ground along the periphery of the Air Force installation. He could spot the buildings, aircraft hangars and Iraqi MiGs parked along the ramps. But there was no sign of the downed pilot. Continuing on the long trip back, Trask avoided the lookout tower as he crossed back over the border into Saudi Arabia and hovered in to land at Ar'ar.

Nobody talked. There was a feeling of disappointment among the Pave Low crew as they remained seated inside the helicopter breathing the gasoline fumes in the hot, stinking refuelling pit. They had been ordered to proceed further south to the main Special Forces base of Al-Jouf. Then, just as the hose was being pulled off their fuel tank, they overheard a brief radio contact between a patrol of A-10 'warthogs' flying over Iraq and the downed pilot. 'We didn't think twice about launching again into a second sortie,' says Trask. 'To debrief another crew would have taken fifteen to twenty minutes. That's valuable time. We already knew the route through Iraqi airspace and were all pumped up with adrenalin. It was the natural thing to do.'

After take-off procedures they immediately lifted off and headed back to Iraq. On receiving map coordinates from the AWAC on where to RV with the A-10s going by the call sign of 'Sandy Lead', Trask entered them into his flight computer. He kept monitoring the A-10s' communications with the survivor at all times but could only hear Sandy Lead's side of the conversation. 'The only way to establish direct communication with the downed pilot was to use our radio direction finding equipment and home in on his survival radio transmission. But the Iraqis could home in as easily as we could so we didn't want to do that until the helo was in a position to make a quick pick-up.'

Fifty miles into Iraq, Moccasin Zero Five made contact with Sandy Lead, who passed on the survivor's coordinates by code. But it was only a rough location. The A-10s had neither maps or the Pave Low's advanced inertial navigation system. There was more bad news. The planes were very low on fuel. If Trask continued towards the survivor, it would have to be without fighter protection until the A-10s refuelled off a KC-135 Air Tanker at 30,000 feet which could take some considerable time. 'If we lose him again, we may never find him,' said Trask and the Pave Low pressed forward alone, deep into denied territory.

There is a physical sensation experienced by a man crossing the barrier of fear. It's not the adrenalin-fuelled excitement of sudden combat or the dizzy elation of jumping off a plane or locking out of a submarine but rather a heavier feeling of pushing through an invisible wall. It requires an individual commitment that is strong enough to break through the pressures of knowing that one is entering, totally alone and vulnerable, a completely hostile environment. Those who live by breaking this barrier have the right stuff of a special operative.

If the new coordinates which he had received from the A-10s were correct, Trask thought, the F-14 had gone down considerably north of the location which he had been given on his earlier run. Maintaining an altitude of twenty feet over the next fifty miles of desert, he received several more calls from the AWAC. Advanced Soviet-made SA-8 and the much-feared French-made Roland missile-guidance systems were scanning his path. Then, stretched over the horizon before him, Trask could see the heavy traffic running along the Baghdad–Jordan highway, marking the midway point of western Iraq.

'I dipped down to ten feet and circled for a few minutes about a mile south of the highway, waiting for a break in the traffic. Many of the passing trucks I saw were military.' As soon as there was a break of a quarter of a mile, Trask pushed the cyclic control forward for speed, coordinating the collective to maintain his altitude, and shot across the road at ten feet. Flying level with the roof of one of the trucks, the Pave Low passed right behind it. 'I'm sure we shocked him but as far as I know, nobody shot at us.'

But Trask now found himself in the clutches of one of the most dangerous weapons systems used against helicopters. The coordinates for the downed pilot given by the A-10 patrol took the Pave Low to within two miles of a Roland SAM site. Specifically engineered to counter low-flying aircraft with its quick-acquisition electronics, the radar-guided missile was considered the number one threat by SAS Chinook helicopter teams, who took great care to avoid their locations when mapping out their flight plans into Iraq.

Straining not to hover an inch higher than ten feet, Trask barely dodged the radar lock which would instantly trigger a missile to turn his helicopter into a burning carcass. Beads of nervous perspiration streamed down his face as he kept a constant eye between the luminescent smudge of the Roland's guidance system showing up on the electronic sensor screen and the horizon closing before him. Manoeuvring the Pave Low to within half a mile of the missile site, searching for the downed pilot, he could see the trailers, barbed wire and sandbag emplacements as 'they kept trying to get a lock on us. We were flying too low to be acquired on their guidance system but they could see us as clearly as we could see them.'

Suddenly, Trask saw a slender trail of flame and smoke appear before his windscreen and heard the Roland streaking a dozen feet above the Pave Low. The Iraqis had made a visual launch. But without a radar lock the missile had overpassed the helo. Trask immediately swung east, manoeuvring with cyclic and pedal controls to fly away from the missile site. Dropping still lower, pushing with his collective down to nine feet, eight feet, then a few inches further to almost seven feet, he felt the cushioning effect of the downwash of the rotors as he almost touched the flat desert coming up beneath him. Its contours were invisible save for on the terrain-following radar; only the digits on the screen showed his altitude. Through the cloud of dust he saw more gravel-peppered sand than blue sky in front of him as he hugged the ground, circling some eight miles from the Roland installation, trying to get out of its range. If some vehicle with a mounted anti-aircraft cannon now turned up and the helo got sandwiched between that and the Roland, he would be dead meat. The minutes passed like hours while he waited for the A-10 patrol to return from their mid-air refuelling.

Descending to 1000 feet, the A-10 fighters re-established radio contact with the downed pilot and Trask could hear them telling Slate 46 to hold down the mike button on his survival radio so that the Pave Low could switch on its direction-finder and pinpoint his location. Getting an immediate bearing on him, the helicopter flew three miles further north and finally picked up the pilot's voice on its radio. Trask knew that the Iraqis would now also be listening in and he had to spot him quickly.

Within the next minute he saw the flash of a signal mirror and caught sight of the survivor hidden in a little hole in the sand. There was less than one mile between them. Trask, already in a hover, trying to avoid the cloud of sand, slowed the Pave Low for a landing. Just then, his left gunner, Sergeant Tim Hardwyck, spotted a green-canvas-covered Iraqi Army truck about half a mile in front, speeding up fast. The A-10s saw it too and Sandy Lead came on the radio, asking, 'What do you want us to do?'

'Smoke the truck,' came the immediate reply from Pave Low co-pilot Lew Caporicci. 'Let's send up another few candidates to Allah.'

Trask banked to the left and accelerated as the A-10s came in for two low passes, firing their 30mm Gatling guns. Stitching up a blanket of dirt and smoke, the first A-10 missed the target. But the second drilled a few dozen of its depleted-uranium shells straight into the truck, which burst into flames just fifty yards from the downed pilot. 'All I could see was a plume of smoke as I turned the helicopter 360 degrees to come back around,' remembers Trask.

'Just fly at the smoke,' came Sandy Lead's urgent instruction over the radio. At about 200 yards, Trask saw the survivor standing up in his hole, getting a clear look at him for the first time. He pushed down the collective and dumped the Pave Low just twenty yards from the burning truck. The inevitable dust cloud which all the crew so despised enveloped them as Hardwyck tried keeping his .50-cal machine-gun firmly trained on the vehicle in case any surviving Iraqis spilled out. The pararescue man cocked his M-16, crouching over the ramp coming down on the ground, flooding the inside of the helicopter with dust.

'When the dust cloud cleared, our survivor was standing just in front of the cockpit, holding his radio up in one hand and a small pack in the other. I cleared the pararescue man off the ramp.' He went around the side of the helicopter to meet the Navy lieutenant sprinting towards him. Meeting him halfway, beneath the .50 machine-gun of the right window gunner, the pararescue man grabbed the survivor and led him up the ramp. 'As soon as he was on, we pulled pitch and took off directly south.' The Pave Low had been on the ground for less than thirty seconds. It was almost exactly 2.00 p.m., six hours since the rescue operation had begun.

'How did you come across the first time,' asked Sandy Lead as Moccasin Zero Five once again approached the Baghdad–Jordan highway. Trask told him. 'Well, we can do it your way or we can just shoot across,' suggested the A-10. There was a good opening in the traffic and the proposed air strafing was not necessary as the Pave Low choppered over the road. Sandy Lead bade farewell and the A-10s streaked up into the skies to refuel from their tanker. Being in a hurry, they hadn't taken much gas the last time.

'We still had over an hour to go to get to the border. But for the first time, we really felt that we were going to pull it off. The Navy pilot walked up to the cockpit and put a hand on mine and Caporicci's shoulders. He said something but couldn't be heard over the noise of the engine and the throbbing blades. We knew what he wanted to say and gave him the thumbs up.' An HC-130 orbiting close to the border came on the radio offering to fly north and refuel the helo. Trask told him that he had just enough to make it to Al-Jouf but added, 'please stand by, over', in case he had to vary his flight path.

They landed among a cheering crowd and the first in line to greet them with a warm hug and a handshake was the base commander, J.V.O. Weaver. Now a lieutenant colonel, Weaver was elated with the successful SOS rescue operation, feeling vindicated twelve years after his tragic experience in Iran. They called him 'The Mayor', as he ran his headquarters from the local town prison, where he held extensive planning sessions with the commanders of Delta and SAS, who also operated out of Al-Jouf.

Trask had pulled off what was to be the only successful search-and-rescue operation of the Gulf War. When a regular Air Force MH-60 tried it some weeks later, the helicopter was shot down. Its crew were captured and the female medic on board was raped by Iraqi soldiers. Captain Trask was awarded the Silver Star, the second-highest US military decoration, and won the Air Force Trophy for 'the most meritorious flight conducted by any pilot that year'. He had flown through enemy radar defences four times, without a break, in one day.

Among the crowd gathering to congratulate Trask were a group of British pilots of the RAF's 7th Helicopter Squadron, also based in Al-Jouf, to support SAS missions behind the lines with their twin-engined Chinooks. 'We are specialist pilots,' says one of them, a fast-speaking, blonde-haired, blue-eyed flight lieutenant. 'I was

personally picked by the Boss to fly SAS missions.' The SAS
individually selects its own pilots from RAF 7th Squadron. 'It takes
about a year to become fully trained in special forces flying. If at any
moment one does not come up to scratch, he is out.' Most of the SAS
pilots in Iraq had had previous desert experience training in Oman.

The SAS Chinooks are equipped with inertial navigational systems
but otherwise lack much of the high-tech gadgetry such as the all-
important terrain-following/terrain-avoidance radar of American
Pave Lows. Their pilots have to fly at low altitudes virtually 'eyeball'.

'Putting it down at night, seeing through NVGs, is the main trick
we have to learn. In Saudi Arabia, it's normal for the pilots to lose
complete sight of the ground from fifty-foot altitudes due to sand
clouds. It was almost a relief when we got into Iraq because the
ground was more solid than the chalk dust of the Saudi desert, which
was the worst thing about flying there.'

Remaining at altitudes between fifty and eighty feet, SAS pilots
avoid the radar of most land-based systems but are still vulnerable to
detection by enemy aircraft and certain specialized missile-guidance
systems such as the Roland. 'That was our main problem in Iraq.
When planning potential missions, we had to have precise intelligence
about the Roland sites to map our flight paths around them. We also
studied available intelligence on enemy anti-aircraft positions, other
missile systems, and troop concentrations.' Without the forward-
looking infrared radar, dropping down to twenty feet would only be
done in extreme cases and usually meant 'holding' at that altitude
rather then dropping lower or continuing to fly as Trask did with his
Pave Low outside the Roland site.

Not a single British or American special forces helicopter was lost
to Iraqi anti-aircraft systems, although many encountered small-arms
fire from the ground troops. One American helicopter which took a
bullet in its fuel tank was rescued by an HC-130 tanker, which
hooked into it, towing the chopper all the way into Saudi Arabia.

'My runs over the Iraqi border ended up not being that much
different than my previous missions in Northern Ireland, I suppose,'
says the SAS pilot. In one case, when Iraqi ground troops were
shooting at his Chinook, 'I first saw the bullets hitting the nose of the
helicopter through my NVGs and the magnified sparks made me
think that the aircraft had caught fire. It was only when I took off the
goggles that I realized that we were being fired upon and took evasive
action' – by diving to a lower altitude. 'It takes you out of their arc of

fire in a lot less time than ascending, which runs the additional risk of getting you locked into an enemy SAM radar.'

'The RAF Chinooks which operated with us were steam-driven,' says one SAS officer who served in the Gulf, 'carrying on under very difficult conditions.' One helicopter struck a land-mine with its back wheel during the 'hot extraction' of one SAS team and despite damage to its tail engine continued to fly, getting the patrol safely back to Al-Jouf. Not disposing of mid-air refuelling and the high-tech refinements of their American counterparts, however, the SAS helicopters could not perform the type of deep, extended missions conducted by Pave Lows. When a search was sent out for the SAS Bravo Two Zero patrol after it became 'disconnected' almost 200 miles inside Iraq, a Pave Low got the job. When a British Jaguar jet on routine patrol over northern Iraq in 1994 developed engine malfunction and its pilot was forced to eject over Kurdistan, an American SOS Pave Low based in Turkey flew in to recover him.

Back at the 'Hootch', a shack standing on oil drums at Hulburt Airfield and fully equipped with outside latrines, the SOS pilots crowd together in their green-khaki flight suits and leather jackets on a Friday night, drinking pitchers of beer. The talk is loud and mostly about helos. Some may mention going to The Rose down the road, where gorgeous girls from all over the south-eastern United States do topless dances for five-dollar tips.

It's the exclusive den of a man's world thick with masculine camaraderie. A collage of scrawled graffiti and memorabilia covers the walls. There are the tail number-plates of a Pave Low which crashed into a sand dune in Saudi Arabia. A sentence in thick black magic marker reads, 'Ghenghis Khan was a liberal.' Another announces, 'Sexual harassment in this area will not be reported but will be graded.' But the one that makes you think, written on the wall leading out to a small porch, perhaps best summarizes special operations flights: 'Have you ever danced with the devil in the moonlight?'

'I can see the curvature of the earth. There are lights below but the landing zone is black. The closer you get the blacker it becomes because it is losing light reflection. It's like falling into limbo.' Sergeant Mick McKintyre of the SAS has just thrown himself out of an aircraft from an altitude of 35,000 feet for a HALO (High

Altitude Low Opening) parachute jump. He is breathing through an oxygen mask. To minimize speed and maximize control he is in a free-fall position with arms and legs bent and slightly extended. Subtle adjustments of hand position equalize airflow so that he won't turn over or go into a spin. 'It's very noisy, with wind rushing around you, which gives the sensation of flying.' But Mick has only to look at the rapidly changing digits on his altimeter to realize that he is dropping like a stone, at a speed of 200 feet per second. 'It feels like my head is about to explode.'

HALO is the ultimate means of clandestine insertion, if not the most dangerous. The delivery aircraft flies above radar cover and when the parachute opens near the ground, it will be indistinguishable from a bird on any radar screen. The High Altitude free-fall was used to insert a patrol of the SAS airborne troop into Iraq during Operation Desert Storm. What their mission was remains shrouded in secrecy. Almost every special operations organization in the world has a unit qualified in the various high-altitude free-fall techniques.

'Clouds seem like they actually have substance to them and I instinctually cringe as I fall into one. Next thing I know, I'm through it with moisture covering my face and goggles.' Mick manoeuvres laterally by pulling one arm in and closing a leg. 'The air feels tangible, I can use it to steer like a glider.' Air turbulence suddenly throws him into a spin and he places his arms against his sides, allowing himself to fall head first before stabilizing into a free-fall position again.

'I've been falling now for just over two minutes. I'm approaching 10,000 feet and my head feels clearer. I begin wondering how much longer it's going to go on for, how much more time until the parachute opens and if . . .' The next half minute seems like an eternity before he hears the buzz of his automatic opening device set at 4000 feet and in the next second, up it goes. 'It's a magical sensation as I'm pulled upwards by the big, double-layered, rectangular canopy unfolding above my head.' The M-T IXX, the latest in military parachutes, allows the HALO master to steer with two toggles which dangle on to his shoulders. Pulling down on one, he glides at an angle towards the drop zone, which he recognizes as it becomes visible beneath him. Gently pulling on both ropes as he approaches the ground, McKintyre slows his fall, touching down at walking speed, without as much as a bump.

When Delta rehearsed for the Iran hostage rescue, it tried a plan of inserting an advance party of four men by HALO to plant landing lights and guide in the C-130s carrying the main force on to the improvised landing strip at Desert One. 'Johnny, Pappy, Rex and Mike exited a C-130 at 12,000 feet into a pitch-black, moonless night, manoeuvred themselves right on to the small runway of an abandoned government facility in Texas, set up the landing lights and brought in the aircraft without a hitch,' recounts a Delta officer. Two members of the Australian SAS airborne troop I saw practising HALO managed to steer their parachutes on to the white line of a cricket pitch.

In a HAHO jump, the parachute opens at 2000 feet from the aircraft, becoming a virtual glider which the parachutist can steer for a horizontal distance of fifteen miles on to a landing area. The technique might be useful for a highly secretive insertion in which a special forces team would fly itself over a national border. When a direct-action mission was being planned against Colonel Gaddafi's chemical weapons plants in Libya, the suggestion was made to HAHO the assault force from a requisitioned airliner flying a regular commercial route over North Africa. Without the plane having to deviate substantially from its normal course, the special forces team would sail down to a staging location fairly close to its target.

Over 50 per cent of the SAS soldiers come from the British Parachute Regiment. Most special forces in the French Army form part of elite parachute units – the 2 REP, 11ème Bataillon Parachutiste de Choc, Regiment Parachutiste d'Infanterie de Marine. Delta Force is largely formed by Rangers and Green Berets, in the second of which airborne training is required. An officer in the Green Berets explains how his free-falling abilities helped him engender a sense of camaraderie with special forces units he had to work with in Latin American countries. 'We'd get together to go parachuting on weekends and then drink and exchange parachute wings. It gave us a sense of shared identity and excitement.'

Gendarmes joining France's GIGN, however, undergo what is usually their first parachute jump. Most police CT units, including Spain's rather earthbound GEO, require their men to become airborne-qualified. 'We are considered elite paramilitary units,' explains Inspector Carrion, 'and have to be at the same level as others.' Some of the bravest and toughest special forces types may experience fear of heights or vertigo and are averse to the idea of throwing themselves out of a perfectly good aircraft. SEALs prefer the water and Ron Yaw had always opted out of high-altitude acrobatics, having had enough with the 1000-foot static-line jump required in the Navy unit. But when he was appointed to command SEAL Team Six he had to do his first high-altitude free-fall jump.

The veteran SEAL officer felt like cursing his predecessor, Richard Marcinko, who had instituted the free-fall requirement for Team Six, when the back ramp of the C-130 opened up and he looked down into 20,000 feet of sheer sky. 'I just gritted my teeth and closed my eyes but once I was out I was loving it . . . it was one of the most thrilling sensations of my life.'

Some HALO specialists can't get enough of it and the sky is no longer the limit. B.N. of the SAS is going for the ultimate 'space jump'. In a five-million-dollar project backed by the USA's National Aeronautics and Space Administration (NASA), he will dive off the gondola of a balloon filled with ten million cubic feet of helium from an altitude of 130,000 feet. In the vacuum of the earth's outer stratosphere, under the unfiltered rays of the sun, higher than any aircraft flies, Bruce will free-fall, accelerating to 800mph as he becomes the first human body to break the speed of sound, dropping through 110,000 feet in four and a half minutes before his parachute opens at 15,000 feet. NASA is interested in testing the idea as a method of recovering astronauts or materials from future space shuttles operating from space stations.

'It will be the first time that a man has gone at supersonic speed,' says Bruce. Previous jumps from altitudes approaching 100,000 feet have been transonic. 'It's a world record.' He will jump in a pressurized space suit which will have to be tested as carefully by NASA specialists as would be done for any other astronaut embarking on a space mission. The smallest tear in the suit can result in Bruce's body being sucked out of it and burned to a cinder. One man on a transonic fall who lifted the visor of his helmet at 50,000 feet had his brains boiled. He is now a vegetable. The multidigital altimeter on which Bruce will be keeping a close watch as he drops from the stratosphere was bought at an auction of Soviet space equipment in Moscow.

'Perhaps I will create a new unit: the SSS, Special Space Service,' jokes the tall, dark, curly-haired HALO enthusiast with characteristic SAS aplomb, as he enjoys a beer at a wine bar in the centre of Hereford. It's Christmas time and he teases the waitress about her

Santa Claus outfit: 'Think of all the children of the world who believe in Father Christmas that you are disillusioning.'

Jumping not quite from outer space, and landing less glamorously in icy sea water, B.N. parachuted with B Squadron into the South Atlantic during the Falklands War in 1982.

10

LIKE A THIEF IN THE NIGHT

'Judging distance in poor light when making a vertical descent into high swells can be difficult. So I decided to jettison my harness without delay. I hit my cape releases and the shoulder straps fell away – but only just in time. I was suddenly immersed in water. At least the harness was free, there was no drag and the canopy did not envelop me. The parachute fell away at an angle, blown away downwind. I struggled into my fins to help me tread water. Then I pulled down the toggle and popped my life-jacket. I watched the dull green canopies of the other troop members, many of them kicking violently out of twists, drift slowly down and hit the water.' Soldier I, along with B.N. and the other members of B Squadron SAS floating in the South Atlantic, barely insulated from near-zero temperatures by dry suits zipped over their uniforms, might have considered themselves lucky. A much hotter insertion had originally been planned for them.

B Squadron was going to do a Tactical Aircraft Landing Operation (TALO) in two blacked-out C-130s directly on to Stanley Airfield, surprising 7000 entrenched Argentine troops and an anti-aircraft artillery battalion. The plan, approved by Michael Rose, was to fight their way to the headquarters of General Menéndez, kidnap the Argentine commander and make their escape into the surrounding hills with their hostage in stolen vehicles. The SAS would hold Menéndez until Argentina agreed to withdraw from the Falkland Islands, which had been invaded at the beginning of April 1982. Give me a blindfolded tight-rope walk in a gale-force wind any time, Soldier I remembers thinking to himself after the numbing briefing at the 'Kremlin', the Intelligence section at SAS headquarters in Hereford. One of the British military commanders during the Falklands campaign remembers the operation as 'one of those plans you just wish would go away'.

'Our main worry when the war started was that we would become logistically overstretched in a drawn-out campaign over the winter,' explains Mike Rose. 'To shortcut this situation, the SAS would score a quick and devastating blow which would decapitate and demoralize the Argentines.' From the start of the Falklands campaign, Lieutenant Colonel Rose was in constant communication with Margaret Thatcher via a direct satellite telephone link to 10 Downing Street. He was as sensitive as any of the top commanders to the political pressures mounting on the Prime Minister as she gambled on the single largest British military expedition since World War II. The SAS, as always, was prepared to go the extra mile and the commander of B Squadron, Major C., had proposed the direct raid on Stanley.

The plan was entering into its execution phase as the men of B Squadron pigged out on American T-bone steaks and got blind drunk on hard liquor at Ascension Island, the midway staging base between the UK and the Falklands. Soldier I recalls how one of his fellow heroes of Princes Gate insisted on drinking down a flaming glass of Drambuie, only to scorch his mouth and spit out the burning alcohol on the table, singeing B.N.'s hand. 'It was a crazy atmosphere. We just wanted to put any thoughts about the immediate future out of our minds.'

'Come on boys, get your kit together. The trucks are outside. We are going for it', came the sergeant major's electrifying screams at 5 a.m., shocking the men out of their alcohol-induced sleep. All the heavy belt-fed GPMGs which the Regiment could spare were passed out to every second man so that they could pour on to the tarmac of Stanley airfield with an incessant hail of deadly 7.62mm fire. Crates of link were torn open for the men to stuff as much of the ammunition as they could into their bergens and belt around their chests. They were issued with newly tested 202 incendiary bombs which could be fired from XM203 breech-loading pump-action grenade launchers. Their faces smeared in black camouflage grease, the SAS 'Rambo' squad mounted the trucks waiting to drive them to the runway and into the open ramps of the two C-130s. Once aboard the Hercules transports in the early morning darkness, they were strapping themselves into the red canvas seats when word suddenly came that the mission had been 'scrapped'.

'Thank fuck for that,' Soldier I said out loud. 'We could have taken Stanley airfield but finally decided that it wasn't worth it,' explains

Mike Rose with a playful, mischievous smile. Sometimes the SAS commander betrays an almost childlike quality.

Major C. remained upset about the decision to scrub his plan all the way to the South Atlantic, where B Squadron parachuted into the water to rendezvous with a British frigate which would deploy them to a location off West Falkland. To add insult to injury, HMS *Andromeda* took its time picking up the men in the freezing ocean, leaving some to float around helplessly for nearly an hour. By the time he boarded the frigate, C. had had enough and 'lost his sense of humour'. Grabbing the watch officer, he held him over the ship's rails, threatening to tear his head off and throw him into the sea unless he quickly got on with collecting the rest of the squadron. Relations became so strained between C. and the ship's captain that when a search had to be requested for a missing pallet containing some vital equipment, one of the troopers had to act as intermediary with the captain, with whom C. was no longer on speaking terms.

'C. was a trigger man,' says one of his subordinates in B Squadron. 'He was a good guy but I can't see him in any other walk of life. I couldn't imagine him making it anywhere else except in the SAS.' B Squadron's commander was known in the regiment as 'Dr C.' after an undercover operation in Namibia where, disguised as a doctor, he rescued the wife and family of the country's president, who were being held hostage in a clinic.

When the Falklands War started, C. had also taken the initiative by trading on some favours he had done his American friends in Delta Force, obtaining in the process some of the latest high-tech military equipment. A Delta major had developed a close personal friendship with C., regarding the B Squadron commander as 'the bravest, the toughest, the wisest, the most gentlemanly soldier I've ever met'. Exhibiting the strong sense of bonding which special forces so often develop, Delta's deputy commander and liaison officer with the SAS considered as 'brothers' his SAS counterparts with whom he had shared selection. They had attended White House lunches together and gone on historical tours of battlefields of the American Revolutionary War. The major was happy to satisfy C.'s request for some of the latest state-of-the-art anti-aircraft, heat-seeking, shoulder-fired Stinger missiles which had just recently been deployed with US special forces.

'Without even thinking about it or asking anyone else, I just gave them a whole bunch of Stingers,' says the major, who personally provided four SAS NCOs stationed at Fort Bragg with a crash course on using the sensitive weapons system. Delta gave no consideration to the intense debate raging inside the Reagan administration at the start of the Falklands conflict between those siding with Britain and some Cabinet officers who wanted to maintain a strict US neutrality and not risk alienating an important Latin American ally.

'As soon as Sergeant Paddy Armstrong had the Stingers crated, he picked up the bergen he always kept packed and was off to join the rest of his squadron staging off Ascension Island. I don't think that he even had time to say goodbye to his wife,' recalls one member of Delta who saw him off.

But tragedy struck Armstrong and his Stinger team when a seagull got sucked into the engine of a helicopter carrying them with a troop from G Squadron on a ship-to-ship transfer and the aircraft crashed into the sea. There were no survivors. To the SAS it was a tragic loss of men and know-how. It is reported that during the rest of the war SAS men went around hunting seagulls to vent their anger. Enough knowledge about the missile had been passed around, however, to enable one SAS trooper to lock his Stinger infrared heat-seeking automatic trigger sight on to an Argentine warplane strafing British forces landing at San Carlos Bay and 'the first Stinger ever to be fired in anger', according to Burruws, 'shot down an Argentine Pucara'. Mike Rose was sure to take the missiles with him when he moved on with D and B Squadrons to Mount Kent.

From the moment that a French-built Exocet missile fired by an Argentine jet fighter sunk the British destroyer HMS *Sheffield* in one of the opening salvoes of the war, the threat posed by the radar-guided missile had become an understandable obsession for the British task force. As the main landing was taking place in San Carlos Bay, another Exocet sunk the auxiliary vessel *Atlantic Conveyor* with its precious cargo of heavy-lift Chinook helicopters. More lucky hits against one or both of the British aircraft carriers or the requisitioned *QEII* ocean liner transporting the main troop reinforcements and the entire British war effort could be sunk.

After a series of direct conversations with Margaret Thatcher over his direct Satcom link to Downing Street, Rose prepared to do something to forestall more surprise Exocet attacks. This highly sensitive operation had to be cleared and supported at the highest level. A top Royal Marine commander with the British task force remembers becoming aware that one of his crack helicopter pilots with special training in Arctic survival and scuba diving, was being detached, along with a Royal Navy Sea King helicopter, for a top-secret mission outside the British total exclusion zone around the Falklands. 'This could only mean something on the Argetine main-land.' The Marine brigadier recommended using a helicopter other than the Sea King, for as the aircraft would be flying outside British radar cover it needed to be equipped with a more advanced inertial navigation system. He was told to mind his own business. Uppermost in the minds of the SAS planners was the helicopter's lift capacity. Without their Chinooks, only the Sea King could carry sufficient fuel to fly 300 miles with a dozen men.

In London, the Chilean naval attaché was being summoned for meetings with top officials in Whitehall. The captain had become the main conduit in secret negotiations between Britain and Chile's military junta under General Pinochet. Among matters being dis-cussed was the donation of a British frigate to the Chilean Navy in return for some assistance with the special operations being planned around the two Argentine airbases of Río Grande and Río Gallegos, which serviced the Exocet-carrying Mirage and Super Etendard jet fighters. Located on Tierra del Fuego, on Argentina's southern tip, both installations were just a little over twenty miles from the Chilean border.

Chile had not only enjoyed a long-standing naval relationship with Britain ever since a British admiral trained and commanded the first Chilean fleet, but had its own territorial problems with its neighbour. Argentine claims to a group of Chilean islands along the Beagle Channel, south of Tierra del Fuego, had on several occasions almost led to a shooting war between the two countries. There was little doubt in the military minds of Chile's rulers about where Argentina would turn next if it got away with its land-grab in the Falklands.

On 20 May 1982, the wreck of a Royal Navy Sea King helicopter was discovered inside Chile near Punta Arenas, about 160 kilometres from the southernmost border with Argentina. The uninjured Marine pilot, Lieutenant Hastings, gave a pantomime press conference claiming that the helicopter had developed engine failure and crashed while on a routine patrol. An official of Chile's Foreign Ministry reports that eight SAS men who came off the Sea King were flown to the Chilean capital, Santiago, where they were put up in good hotels, treated to nights out on the town and given cases of Chilean wine to take back to England.

Some newspaper accounts claim that a British special forces team clandestinely penetrated an Argentine airbase to tamper with the high-explosive warheads of stored Exocets, rendering them ineffective, and got out through Chile. A likelier account of the SAS mission was that they planted special electronic sensors on the Argentine airfields to monitor radar frequencies and identify aircraft take-offs and landings. The signals could in turn be picked up at safe distances, either from the Chilean border or from submarines stationed offshore.

A member of the Special Boat Service (SBS) of the Royal Marines claims to have participated in a submarine insertion on to Argentina's coast to set up a hidden observation post near the Comodora Rivadavia airbase. From some high rocks on a cold, windswept beach, he and his four-man team operated for a week, observing aircraft coming and going from the Río Grande area in Tierra del Fuego. During their extraction, the SBS team were careful not to even leave floating traces of their presence, lashing their collapsed Gemini inflatable boat and other used equipment to the submarine's outer casings. Once the vessel had submerged to the lower depths, a diver swam out to slash the ropes holding the wasted gear, letting the stuff sink to the bottom of the South Atlantic. When the nuclear submarine HMS *Conqueror*, which sank the Argentine cruiser *General Belgrano*, steamed back into port after the war, a dagger was painted beneath the Jolly Roger flying over its bridge. The symbol meant that the vessel had engaged in an SBS insertion.

The British were never sure about how many Exocets Argentina had in its arsenal, although the number was not believed to be substantial. The intelligence service MI6 did know that the Argentines were desperately trying to acquire more of the missiles through their embassies. The kind of money that the military junta in Buenos Aires was willing to put up ensured that there were plenty of arms dealers lining up to broker the deal. If Argentina was able to secure an unlimited supply of Exocets and launch a massed missile attack on the British fleet the effects could be devastating. Major C., who was also the SAS's chief intelligence officer, came up with another of his ideas – to infiltrate a covert special forces team into France to blow up the factory producing the much-feared missiles.

Sources at SAS headquarters in Hereford remember C.'s suggested raid into France being seriously discussed among units being held in reserve back in Britain. But unwilling to risk a war with France, the British Foreign Office shot down the idea. It is not certain if the matter was ever discussed between Lieutenant Colonel Rose and Prime Minister Thatcher. Some believe that if it had been, 'Maggie would have gone for it'. But as things stood, another one of C.'s plans had been scrapped. It just wasn't his war.

As British forces began their slow advance inland from their beachhead at San Carlos Bay, the Commander of Land Operations, Brigadier Julian Thompson of the Royal Marines, had to take about two hours out of his busy day to helicopter back and forth to the SAS satellite communications room, the only one of its kind in the British task force. The British government wanted daily progress reports, pressuring Thompson to move faster against Port Stanley. But the cautious and methodical Marine commander would not be rushed, bedevilled as he was by a fragile supply line and a lack of mobility resulting from the sinking of his heavy-lift helicopters by the Exocet attack on their transport ship.

Listening in to the strained discussions between Brigadier Thompson and 10 Downing Street, Rose became acutely aware that Mrs Thatcher wanted a quick and dramatic demonstration of a British military breakthrough. His realization that this was a political necessity from the start of the war was the reason why Rose had at first supported the launching of a lightning raid on Stanley airfield. The man who had given Thatcher her SAS victory at Princes Gate now resolved to do the same in the Falklands, even at a time when the outcome of the South Atlantic campaign looked highly uncertain.

An SAS reconnaissance patrol which Rose had sent in to observe Argentine positions around Port Stanley weeks before the main British landing had been working miracles. Captain Spalding's team, operating for sixteen days without resupply, had directed an air strike which destroyed most of the Argentine helicopter fleet, denying General Menéndez a critical advantage which he would have otherwise held over the British. The enemy commander could no longer move sufficient reinforcements forward to effectively block the main advance spearheaded by paratroopers and Royal Marines.

Captain Spalding had climbed to the top of Mount Kent, the highest peak around Port Stanley, to discover that the strategic high ground was virtually unoccupied. The position dominated most

Argentine defences and was only ten miles from Menéndez's head-quarters. 'Mount Kent is the key ground in the Falklands,' concluded Rose, who proceeded to seize it. On 24 May he scraped together two helicopters to airlift D Squadron, reinforced by an element of B Squadron, to the mountain top, where they met up with Spalding's eight men already camped there. 'I turned a small hidden OP forty miles ahead of the British front lines into a fire base,' says Rose. GPMGs (General Purpose Machine Guns), 81mm mortars, 66mm rockets and Stinger anti-aircraft missiles were set up in defensive positions among the boulders and craggy rocks.

Using the media in a way in which no SAS commander had previously done, Rose flew up to Mount Kent accompanied by a press pool consisting of correspondents from the BBC, *The Times* and the *Daily Telegraph*, so that the world could see that British forces were at the gates of Port Stanley. The reporters had agreed in advance not to reveal that it was an SAS operation. Only Max Hastings of the *Telegraph* hinted at it with a supposed quote in which he had Rose saying, 'Who dares wins.'

It was a dangerous piece of theatre which engaged the SAS in some of the toughest combat of the war, costing the Regiment its first serious casualties from enemy fire. Argentine C-130s overflying the position at night on their regular runs into Port Stanley, rolled out 1000lb bombs from their back ramps on to Mount Kent. One soldier on a reconnaissance near the base of the mountain with SAS veteran 'Hector the Corrector' of Oman and Princes Gate fame recalls thinking at first that they were under a directed artillery attack. 'We took it very personally, thinking that they had spotted our patrol and the huge explosions which shook the ground like an earthquake were aimed specifically at us.'

'Hector, are you scared?' he asked as they cowered for cover behind rocks. 'Damn right I'm scared,' came the honest reply from a man who had a reputation for thriving on every fire-fight he had ever been involved in. It was not until the morning that they realized that the bombardment had come from the air, seeing the shrapnel-filled ten-foot craters leading all the way into Port Stanley.

'We were under intense harassment,' says Rose, 'but missing the Chinook helicopters which we had taken out, Menéndez could not react with a large enough force to overwhelm us on Mount Kent. For the first couple of days we encountered a few Argentine fighting patrols which were easily beaten off.'

But the British lack of helicopter mobility was also delaying plans which had been agreed with Thompson to reinforce the Mount Kent position with a battalion of 42 Marine Commando, and during his third morning on the peak Rose observed something which sent a chill up his spine. Through the high-powered telescope binoculars which the SAS had trained on Stanley airfield, he could see a large group of soldiers unloading from a C-130 who looked distinctly tougher and more professional than the ill-trained and undisciplined conscripts the British had mainly encountered thus far. They wore obdurate expressions under their black wool watch caps and carried a lot of high-quality equipment, including GPMGs and 'American-made third-generation image intensifiers which were as good if not better than anything we had for night vision'.

Rose was watching Argentina's elite 601 Commando Force, schooled in special operations by American Green Berets. Many had trained at Fort Bragg and were hardened in the ruthless counter-insurgency war which Argentina's military junta had un-leashed against Cuban-supported guerrilla organizations. Some Argentine commando officers were serving under contract with the CIA in Central America training the right-wing Nicaraguan Contra guerrillas. There was no doubt about what they had been brought to do in the Falklands. The battle for Mount Kent had become a contest between rival special forces. The SAS braced itself for a match.

The cold South Atlantic wind drowned out the throbbing engines of the two remaining Argentine Chinooks landing on low ground near Mount Kent under the cover of darkness. The forty well-camouflaged commandos who emerged divided up into three teams and crept slowly towards the base of the mountain, stopping, listening, obser-ving the green and black images in their NVGs to pick out any heat sources. The point man of an SAS patrol moving along the lower slope failed to notice the half-dozen Argentines who had spotted his smudgy silhouette. Slithering through waist-high, soaking wet peat, half the commandos crawled around the flank of the British patrol for an L-shaped ambush. When they were within twenty yards of the patrol they opened fire. The SAS man was immediately struck down by 7.62mm rounds hitting his stomach and chest, coughing up blood as he fell. The three others spaced out some distance behind him and dived for cover, returning fire in all directions as they became enveloped by a barrage of automatic arms fire and grenades which injured another SAS man with shrapnel. He had to be partially carried by the other two survivors as they withdrew up the mountain.

Another SAS patrol on a defensive position further up the hillside managed to hold back the pursuing Argentines, covering the ambushed group's escape. But as the forty-man Argentine commando force consolidated and concentrated their fire, 'they closed our position', according to Mike Rose.

The six SAS men, one bleeding severely from the metal embedded in his leg, spent the rest of the night moving as fast as they could up the mountain, stopping intermittently to lay suppressive fire on the Argentines closing behind them until the men of 601 Commando were checked by mortar fire from SAS positions near the peak.

The following day became a cat-and-mouse game as the Argentine commandos kept probing the SAS positions with sniper fire and occasional 60mm mortar rounds while managing to remain sufficiently concealed from British efforts to pinpoint their locations. As darkness fell the fighting intensified and two commando patrols, about twenty men, reached the top of Mount Kent. 'They got within grenade throwing distance of my headquarters,' reports Rose, who picked up his Armalite and joined the fight as shrapnel flew all around him.

It was a confusing intermittent night battle, fought mostly at close quarters with running shadows, sudden muzzle flashes from automatic weapons, then silence. A blackened enemy face would appear unexpectedly from behind a rock just feet away, and shots were exchanged, followed by a grenade exploding. 'The Argentines liked to use grenades and had lots of them,' says Rose. Because of the confused battle lines, the British mostly refrained from using their rockets. The SAS finally managed to surround the main commando group, consolidating into a position near the peak, and ambushed them with one of those devastating, explosive onslaughts of automatic rifle and GPMG fire for which the Regiment is famous.

The survivors hastily withdrew, leaving behind two dead and six wounded, who were taken prisoner. 'By the end of the night we had whittled them down by about half,' says Rose. Two more men were wounded, but the SAS remained in control of its main positions by the morning of 30 May.

Elements of 601 Commando kept up harassment actions around Mount Kent as British Marine and Parachute Regiment reinforcements began arriving over the next few days. 'There was one Argie sniper in particular who gave us a lot of trouble,' says a soldier from B Squadron. 'He was very good at camouflage and concealment and it took us almost a week to hunt him down. During that time he took out thirteen Paras.'

Using Regimental scout telescopes with 20x magnification, three four-man patrols were constantly stalking the elusive sniper, forming arcs of observation around the various points where his shots were heard. 'The worst part of the operation was the weather. It was just awful being out all the time in gusting wind and freezing rain. There was also fog, which made it almost impossible to pinpoint his location, especially as the rifle he used was equipped with a flash suppressor. He kept moving his position at night.'

During the sniper's sixth morning shoot, when one of his rounds hit the radio operator of a para unit moving up Mount Kent, one of the OPs was finally able to observe a puff of smoke coming from an elevated area about 150 yards away. The sniper's exact position was calculated with a prismatic compass and map, and the other patrols were guided by radio to close in from separate angles. Once they had the area encircled from three positions the SAS triangulated single shots from their Armalites. After the first volley they could only see some barely distinguishable movement in what would otherwise have been just a clump of peat. Then a rifle was tossed into the air. The SAS held their fire as a human figure suddenly appeared from beneath the vegetation, covered in hessian cloth and smeared with camouflage paint. He had his arms raised.

They examined the sniper's weapon. It was an advanced American-made high-powered magnum rifle fitted with a laser rangefinder, which explained how almost all his shots had gained their target from considerable distances. The SAS took the rifle back to Hereford with them. Their prisoner they treated with the utmost consideration, offering him food, hot coffee and even a shot of prized regimental whisky to warm him up. 'He had really made us work hard and we respected him for it,' says a soldier from B Squadron. 'It was a gentleman's game. When they fought back we liked it.'

The final blow against Argentine commandos blocking the road to Mount Kent was actually delivered by a platoon of Royal Marines commanded by Captain Rod Boswell of the Arctic Warfare cadre, who had once failed SAS selection. Earlier in the campaign, Boswell had got into a scrape with Rose when he was tasked to cross a sea inlet by small boat to attack an Argentine position. Knowing that the

SAS had an OP operating in the area, he had asked Rose if it was possible to liaise with his patrol to gain further intelligence on the enemy positions. 'It's none of your business,' snapped the SAS commander, exhibiting the zealous attitude he had displayed throughout the campaign of keeping all SAS information tightly compartmentalized. 'Up to the level of brigade commander it was normal to keep other units unaware of special operations taking place in their area,' explains Rose.

Boswell claims that the SAS commander also did not want to admit that he had lost radio contact with his OP. Unable to launch his rigid raider inflatable boats through the heavy surf, Boswell and his Marines returned to headquarters soaking wet and covered with mud, to be derided by Rose for 'masquerading as an Arctic warfare group'.

On the morning of 2 June it was one of Boswell's own OPs which spotted about thirty Argentine commandos taking refuge from the freezing weather in Top Malo House, a small shelter built for shepherds on a plain beneath the mountains around Port Stanley. Overcast skies obstructed an air strike by any of the seventeen overtaxed Harrier jets providing the only air cover for the entire British task force. Boswell immediately set about finding a helicopter which would fly his force from San Carlos Bay to attack Top Malo. A cheerful, bearded, overweight pipe-smoking Navy pilot volunteered his small Wessex, into which Boswell piled in with twenty men. 'Lying practically on top of each other, we made the entire trip flying at no more than ten feet off the ground and landed a few hundred metres from the Argentine position'.

Boswell divided his platoon in two groups. A fire-support team moved to take positions on high ground just south-east of the building, while Boswell led the main assault team of a dozen men crawling several hundred yards through the soggy peat and slushy snow until they got within fifty yards of the house. 'We realized that our slow, wet, torturous approach had been unnecessary as there were no sentries posted outside.' The captain immediately ordered one of his Marines to fire on the house with M-72 rockets. The first fell short, the next destroyed the porch and the third impacted on the smoking chimney, blowing off the roof.

The commandos poured out of the burning house, laying down thick automatic fire, pinning down the Marines behind a fence. 'They were tough and their firing was accurate and very intense. We could not advance any further.' Within minutes, two of Boswell's men had been wounded. When one of the Marines threw a grenade which exploded close to an Argentine, 'he just got really pissed off and came at us hosing down bullets from his automatic rifle like crazy. He was very serious.' The Argentine hit another Marine in the shoulder before Boswell could take precise aim and shoot him.

Hidden behind the crest of high ground, the fire-support team started firing their 40mm grenade launchers at vertical angles, like mortars. 'It had the effect of making the Argentines believe that they were under bombardment from some vastly superior force which had them surrounded.' After about forty-five minutes of fighting, the Argentines finally gave up. Five of them, including their commander, had been killed. Nine were wounded. Thirteen commandos threw the bolts out of their rifles and raised their arms. Boswell took the surrender from the group's second in command, a lieutenant of Welsh origin by the name of Brown. With their main force taken out, three additional enemy commando patrols gave themselves up the following day.

The Falklands campaign was a more closely-run thing than it was perceived to be at the time of the British victory. When the Paras fixed bayonets to storm Argentine positions at the top of Mount Longdon, they found themselves running into vast minefields. They counted some 1500 anti-personnel mines sown along the slopes, 'but only two exploded' recalls Corporal Peter Cuxson of B Company, who took an Argentine machine-gun position that day 'because the rest were frozen over by ice'. Otherwise 'the final battle for Port Stanley would have been an altogether different story' concludes the ex-Para, who went on to become a Foreign Legion commando with France's 2 REP. By the time the hilltop defences around Port Stanley had been cleared, front-line British units were down to six rounds of ammunition per man.

From his forward regimental headquarters on Mount Kent, Rose was aware of the mounting difficulties and the terrible human cost which would be paid if a protracted battle developed for the town of Port Stanley. Deciding to try more special forces guile and bluff, he used the unique position he occupied to open up his own radio channel to General Menéndez's headquarters. 'On 8 June, I made contact with the Argentine headquarters staff in Port Stanley in order to apply psychological pressure to bring about an early surrender. The British now occupied all the high ground surrounding Port

Stanley and it was important that the battle for the Falkland Islands did not extend into the town. It was worth trying to get General Menéndez to accept an early surrender before such a battle became necessary.

'Although no assurances were given by the Argentine Staff officer, Captain Melbourne Hussey, with whom initial discussions took place, it was agreed that a line would be kept open between our two headquarters. Each succeeding day as the Argentine position became more hopeless greater psychological pressure was put on the Argentine headquarters via the open line.' While offering a carrot, Rose also sent in SAS patrols to apply the stick.

'A patrol from G Squadron managed to infiltrate through enemy lines and establish themselves on Seal Point. They were able to call down observed artillery fire from 4 Field Regiment Royal Artillery accurately upon troop movements and concentrations on the rear of enemy positions. So great was the carnage, that on the morning of 14 June when Argentine forces were fleeing from their positions, a voluntary halt was called by the patrol to the devastating artillery fire on the retreating enemy.' At 1100 hours that same morning, General Menéndez's headquarters contacted Rose through their established radio channel and invited him to lead a negotiating team into Port Stanley to discuss terms for a ceasefire.

Rose landed by helicopter in the middle of town with his radio operator and a Spanish-speaking Royal Marine, Captain Rod Bell, who had been serving as SAS interpreter. Hacking their way through the hedges of several back gardens adjoining the brightly coloured houses, they reached the doors of the school serving as Argentine headquarters. Receiving them with a crisp military salute was General Menéndez, the man Rose had planned to kidnap.

During the six hours of negotiations with Menéndez and his staff, Rose kept eyeing a silver statuette decorating the conference room which depicted the head of Argentina's military junta, General Galtieri, heroically riding a charging stallion. It was a gift inscribed to Menéndez which Galtieri had sent on the day that the Falklands were occupied. 'I want that,' Rose whispered to his radio operator, and when the meeting was adjourned the soldier faithfully followed his commander's instructions and stuffed the statue inside his Gore-tex field jacket. The silver sculpture of Argentina's last man on horseback has since decorated the SAS Officers' Mess in Hereford.

Three Agusta 109 helicopters captured from the Argentines were also taken by the SAS to serve as their private air force.

While sheltering inside a large hospital tent near Mount Kent one morning as scattered duels continued with Argentine commandos, Boswell waved a geeting at the comouflage-clad figure with long, scraggly, red hair who suddenly appeared out of the freezing rain. he was covered in peat, and looked wet, dirty and unmistakably tired. His rifle was filthy. It was obvious that he had been out on patrol all night. 'Hello Cedric, it looks like you've been putting down some rounds,' said Boswell.

'Yeah, we've put down a couple,' replied Major Delves as he sat down, pulled the bolt out of his rifle and started to clean the weapon. The impassive commander of D Squadron was as well known for his brief understatements as for his casual, usually scruffy appearance. Back at Hereford his shirts were always crumpled, with the tail generally hanging out, and adding a touch of eccentric absent-mindedness, after bicycling to dinner parties he would often neglect to pull his trousers out of his socks even after sitting down to eat.

Cedric Delves had had his ups and downs during the Falklands War. He had started on a sour note by sending his squadron's Mountain Troop on an unnecessary and nearly disastrous mission to the top of Fortuna Glacier on the Antarctic island of South Georgia during the opening days of the campaign. Wrinkling up his chin, he had given one of his customary monosyllabic dismissive replies to an experienced Royal Marine Arctic Warfare instructor who advised his against launching the operation. Explaining all about atabatic and katabatic wind systems, Major Guy Sheridan had warned Delves that heat reflected from the ice would rise during the day and blow back down at night with gale-force winds of 100 mph.

No Marine knowing some fancy jargon was going to convince Cedric Delves to call off his approach of the Argentines' rear and the day after Mountain Troop was dropped off on the steep, ice-covered 1800-foot glacier it had to be evacuated. Struggling against the wind as they moved over crevasses and deep snow in temperatures of thirty to forty degrees below zero, the twenty SAS men had been barely able to gain 500 feet during the night. Sleeping without proper shelter as their tents had been blown off the ground, many were suffereing frostbite and in danger of hypothermia by morning. The grease in most weapons had frozen, rendering them useless, and as Wessex helocopters hovered in with a very worried Delves on board to rescue Mountain Troop, two of them crashed in the blinding blizzard. It was only due to the daring and nearly miraculous flying of one Royal Navy pilot that the SAS remnants were pulled off Fortuna in two separate runs between the galcier and a warship offshore.

As it turned out, just a few rounds of naval shelling on the small settlement of Grytviken was all it took to convince the Argentine garrison of 170 men huddled inside their headquarters to put up the white flag. When D Squadron landed with a detachment of Marines, the only shooting was against a herd of unfortunate walruses whose silhouettes were mistaken for the hooded figures of Argentine soldiers.

It was an inauspicious beginning. But Delves remained unflappable. The near-fiasco on Fortuna Glacier – described in Margaret Thatcher's memoirs as one of her worst moments of the Falklands campaign – may, in a strange way, even have vindicated some of Delves's deeply held beliefs. He was of the attitude that while human foibles are trivial and fleeting, nature is permanent. It was the forces of nature which had overcome his Mountain Troop and it was the human enemy who later surrendered to his squadron's sergeant major without firing a shot. An officer who did not usually wear any sign of rank and tolerated the most unruly behaviour by the Paras who composed D Squadron, Delves had once severely reprimanded a soldier for snapping a young branch from a tree. He loved both nature and his garden.

Having transferred with his squadron to the aircraft carrier *Hermes*, the flagship of the British Task Force, Delves was soon in Admiral Sandy Woodward's operations room in front of a large map, pointing the dagger which he always carried slung to his belt at the red circle marking the Argentine airfield on Pebble Island. Based there was a squadron of propeller-driven Pucara attack planes within easy striking range of the British landing point in San Carlos Sound. Instead of risking the aircraft carrier's precious few Harriers on an air strike which was unlikely to achieve the objective of destroying all the enemy planes, Delves suggested landing Mountain Troop to neutralize the airfield. Finally convinced, Woodward gave the SAS five days to carry out the task.

Before getting the Task Force commander's final consent, Delves inserted a patrol on Pebble Island by canoe for a recce of the airfield. Crawling up at night with three of his men to within 2000 metres of the installation, Captain Tim Burles observed a total of eleven aircraft well dispersed as a measure against air attack. There was a large, well-entrenched garrison constantly patrolling the perimeter. As it was flat, open country with no hills and little vegetation in which to take cover, Burles decided to move out with his patrol as dawn broke. They left their bergens buried in the ground to enable them to slither away quickly to the east without being spotted. After linking up with the four men who had remained behind with the canoes in a beachside area, the patrol leader immediately encrypted his intelligence on the key pad of his Plessey field radio. On board the *Hermes*, Delves received the short burst transmission and began to prepare his plan.

Four nights later three Sea King helicopters carrying D Squadron took off from the aircraft carrier. The SAS men were armed to the teeth, as B Squadron had been for the cancelled raid on Stanley airfield. All Armalites were fitted with M203 pump-action 40mm grenade launchers. They also carried GPMGs, demolitions charges and 66mm rockets on disposable launchers. The black silhouettes of the helicopters approaching without lights at low level were barely discernable against the dark-blue sky before the rotor blades became audible to the Argentine troops on the ground. A barrage of naval gunfire exploded among the enemy trenches as two Sea Kings landed on the airfield. The men were pouring out before the aircraft touched the ground, blazing away with their rockets, grenades and automatic fire. Several stationary Pucaras instantly exploded into flames.

The third helicopter landed outside the airfield's perimeter, unloading the fire-support team, which linked up with Burles's patrol who lay waiting there. They set up machine-gun positions and mortars to cover the assault force and cut off possible routes of enemy reinforcement.

The night sky suddenly erupted into yellow and pink as Argentine ammunition and fuel stores exploded and the men of Mountain Troop fanned out through the airfield to complete the destruction of the scattered warplanes. They systematically planted charges inside the air intakes beneath the wings, as David Stirling's men had done against Rommel's air force in the North African desert forty years before. But the Argentine troops managed to regain their balance and started to counter-attack with machine-gun and grenade fire as a lone trooper ran back from a far corner of the airfield, where he had just blown up the last parked Pucara. In the process of covering him another SAS man was wounded by shrapnel.

The assault force began withdrawing, gathering on the airfield's outer perimeter, where a remote-controlled land-mine exploded in their midst. Two men flew into the air, one of them suffering a severe brain concussion. The helicopters landed for the pick-up and Burles's patrol piled into the first one, followed by the fire-support team. The assault force got into the other chopper two minutes later as one man providing rear cover pumped off a stream of grenade rounds from his M203 launcher, cutting down a group of pursuing Argentines. All the SAS raiding force lifted off back to the *Hermes* as dawn was breaking, leaving behind eleven destroyed aircraft and a burning airfield. The attack had lasted less than thirty minutes.

Cedric Delves had directed what was perhaps the most dramatically successful SAS direct-action raid since World War II at the cost of only two wounded. Rising from near failure at Fortuna Glacier, he had turned the Falklands into his war.

11

IN THE VIPERS' NEST

'We walked all night, never quite knowing when we had crossed the border. I only realized where I was when I got to the top of a ridge and saw a Toyota truck through my binoculars with Libyan markings and some Arab-looking soldiers riding in the back.' The last time Peter Cuxson had been in action was as a British Para in the freezing South Atlantic. He was now with the Commando de Recherches et d'Action dans la Profondeur (CRAP) of the French Foreign Legion's 2 REP on a top-secret mission in the scorching deserts of Libya. They were lightly armed, carrying 5.56mm FAMAS CAR-15 rifles with attached M203 grenade launchers. Dark glasses and Foreign Legion desert *shemags* covered their faces.

It was June 1984 and the 2 REP had been deployed in Chad for Operation Manta, ostensibly a defensive action to protect Libya's southern neighbour from Gaddafi's incursions. US special forces teams, including elements of Delta, were also on the scene training illiterate troops of Chad's President Hissène Habré in sophisticated weapons systems, such as the Stinger, to shoot down Libyan aircraft.

Operation Manta was being secretly expanded to include surveillance missions and possible direct-action operations against terrorist training bases north of the 16th Parallel, which divided Libya from Chad. The camps were being used, according to the intelligence briefing received by Cuxson and the rest of CRAP, by extremist Palestinian guerrilla factions and some European terrorist organizations. Satellite reconnaissance had identified the facilities and French Intelligence, the Direction Générale de Securité de L'Etat (DGSE), wanted close ground observation of the camps. Photographs were even handed out of certain individuals CRAP should look for. Divided into six surveillance teams, the Legionnaires were given instructions to report back to headquarters immediately any information they gathered. The separate patrols were dropped off by jeep at several points along the Libyan border as night fell over the desert.

By morning Cuxson's four-man team had covered about half the distance to their objective. With the whip lash of the North African sun beginning to bear down, the Legionnaires stopped for the day. After digging shallow scrapes in the ground, each man piled bits of vegetation and dirt over their hessian cloths and pulled them over their heads as they crawled into their holes. They lay there, keeping still until sundown. When they resumed their march on the second night, Cuxson walked in the lead, about 200 metres ahead of the rest. Suddenly he caught heat sources at some distance. Using infrared binoculars, he peered at six buildings surrounded by a wire fence, about a kilometre away. He could detect some human movement inside the compound. Checking his map references, he confirmed the location of the camp which his patrol had been sent to observe.

It was barely an hour before daybreak as Cuxson and the other three men crawled up to a position on slightly elevated ground 500 metres from the camp's perimeter. There they dug scrapes and settled into them. It was 6 a.m. when they began to hear distinct sounds coming from inside the camp. Two men trained their binoculars through small openings in their netting while Cuxson started snapping photographs with a camera fitted with an 1100mm telephoto lens. The radio operator punched the key pad of his radio set, every few minutes sending out burst transmissions reporting their observations.

They estimated that there were between thirty and forty people inside the compound. As the morning wore on, it was possible to distinguish seven or eight Libyan military instructors, a small sentry force and support personnel. About twenty trainees were being put through firearms drills after their PE session. During the next few hours the CRAP team observed them shooting at cardboard targets with Kalashnikov AK-47 rifles from various firing positions, after which they did an hour of pistol practice.

After walking some distance to a wide, open area littered with corrugated-iron debris and broken-down vehicles, the trainees took turns firing Soviet-made RPG-17 anti-tank rocket grenades. The instructors next brought out a highly advanced German-made Armbust portable missile. Equivalent to an American Light Anti-Tank Weapon (LAW) with disposable launcher but with greater hitting power and no flash or heat signature, it is the ideal terrorist weapon in an urban environment. It can be fired from fairly close range into a busy city street without anyone being able to tell its firing point or trajectory.

Observing the layout of the camp, the CRAP team reported some half-dozen vehicles, including Range Rovers and Toyota trucks parked in a covered motor pool. Among the six single-storey buildings, they were able to identify the radio hut and headquarters area, two separate long concrete blocks used as living quarters and another similar structure which seemed to serve as a classroom and communal facility. There was also an arsenal and storage depot.

After lights out in the camp at 9 p.m., Cuxson scanned its perimeter intently for nearly an hour through his infrared binoculars. No mobile sentry patrol could be observed. He and another Legionnaire attached laser sights to their rifles, covering the two other men, who lifted the camouflaged netting and crept out of the hide, slithering through the dirt like snakes for 250 metres until they reached the perimeter's barbed wire. Digging small holes in the ground, they planted two IREMBASS (Improved Remotely Monitored Battlefield Sensor System) sensors so that future surveillance teams could monitor activity in the camp from safe distances.

Once the pair had crawled back, all four men immediately started vacating the recce site. Taking extreme precautions not to leave any traces, they scooped the dirt back into the hole, padded it down neatly and took their rubbish back with them. Moving to a marked position about a kilometre away, they dug up the bergens and extra gallon tank of water stashed there and radioed for helicopter extraction. They marched over ten kilometres that night for the prearranged RV with an American M-60 Blackhawk.

Cuxson spoke with members of other CRAP surveillance teams returning from their respective operations. One patrol reported coming across a large compound with about twenty buildings but finding it mostly deserted but for a handful of Libyan military personnel. Another four Legionnaires talked about feeling very scared as they lay camouflaged outside an extensive training facility with big two-storey buildings and nearly one hundred people undergoing intensive demolitions and firearms training. They observed them practising assault techniques and wiring car bombs. It was one of the bases which was specified for further intelligence gathering and the planting of special receivers to guide in missile strikes.

During the next two weeks CRAP activities intensified inside

Libya. Further recce patrols were conducted around two of the terrorist camps and sabotage operations got under way against infrastructure targets.

It was about six weeks since Cuxson had surveyed the Libyan terrorist camp and lights were going out at the usual time, between 9 and 10 p.m. The two sentries at the guardpost paid little attention at first to the whirring noise of the three helicopters materializing out of the night sky. It was exactly midnight as two of the Blackhawks hovered over them at an altitude of twenty feet and their landing lights illuminated the whirlwind of dust kicked up as the machines touched down inside the compound. With all the noise it was difficult to tell if the sentries shot off some AK-47 rounds before their small post exploded with the impact of LAW rockets and both men were hosed down with machine-gun fire.

Twelve silhouettes streamed out of each helicopter, running towards the main buildings in two single files. One group immediately made for the headquarters and radio hut. Eight men entered the buildings as four pumped incendiary rockets into the nearby motor pool, exploding all six vehicles into flames. The other dozen men surrounded the living quarters, pouring grenades and thousands of automatic rounds into each building before entering to finish the job of killing everyone inside. Dead bodies were carefully counted along the cleared corridors and rooms, and videoed close up for identification purposes.

The headquarters and radio installation disintegrated under the exploding demolitions charges as half the raiding force piled back into their Blackhawk, which they loaded with files and communications equipment retrieved from the razed buildings. Next the arsenal and storage depot burst into flames, sending a fireworks display of stored ammunition and rockets shooting into the air. The rest of the assault force returned to their helicopter and both MH-60s took off while a third Blackhawk picked up the CRAP reconnaissance team from its position outside the camp's perimeter.

Back at 2 REP headquarters in Ndjamena, Cuxson examined a British-made Plessey radio which had been among the equipment seized at the destroyed terrorist camp. The unit which carried out the raid, he maintains, was partially drawn from the 2 REP although operating outside the Legion's direct command. He is also of the impression that US helicopter crews and elements of Delta Force participated in the operation. In April 1986, USAF F-111 bombers were launched from bases in England to conduct air strikes against Colonel Gaddafi's headquarters complex and terrorist training camps around Tripoli. Although France affected not to be involved in the raid, informed sources maintain that French operatives provided on-ground reconnaissance for the air bombardment.

In 1981 the commander of the 2 REP, Colonel Georges Grillot, had been appointed to direct the DGSE's 29ème Service d'Action (29 SA), the secret special operations branch of French military intelligence. His brief was to greatly upgrade 29 SA's counter-terrorist subdivision. 'He brought in a lot of his officers from the 2 REP to undergo intensive three-month courses set up at facilities in Aspretto,' says a top DGSE official. Further training exercises were conducted at the Centre d'Entraînement des Réservistes Parachutistes, in Cercottes. Men from other elite French parachute units and the Navy's Nageurs de Combat were also recruited into 29 SA. 'It took six to seven months to build up our counter-terrorist capability.' Specializing in CQB techniques with sub-machine-guns and pistols, 29 SA operatives cross-trained in all aspects of water and airborne operations, intelligence-gathering, clandestine communications, covert methods of reconnaissance and surveillance work, sabotage and explosives.

'We selected the men individually,' says the DGSE officer, 'looking closely at their military and combat records to determine their courage, imaginative capacity and ability to work in isolation and clandestinity.' Even so there was a 50-60 per cent attrition rate during the training period as 29 SA expanded to a force of 150 men based at a new headquarters in Fort Nuissy, 'where they practise constantly'. Small teams would be deployed from there for high-risk special missions all over the world. 'They often operate in groups of only two men.'

While American Delta men and Green Berets were training the Afghan Mujahidin to fight Soviet invaders at camps in Pakistan, 29 SA teams were operating with the resistance fighters inside Afghanistan. On 7 July 1988, the Soviet news agency Tass reported the killing of two French advisers in an ambush by members of the Soviet Army Spetsnaz on a group of Mujahidin. 'Two-man teams operated with the Mujahidin, being rotated for three-month tours inside Afghanistan,' according to DGSE sources. 'Their main mission was to train handpicked groups of resistance fighters in CQB tactics with French,

Israeli and Russian sub-machine-guns to lead high-level ambushes and assassination missions against Soviet commanders.'

Among its many operations, 29 SA had also sabotaged a pro-Libyan coup in the Central African Republic, where French operatives charged with protecting President Kalinga had detected well in advance preparations for an armed uprising in 1982. Reinforced by a team of half a dozen men sent from France, the 29 SA team quietly organized a local reaction force and seized the broadcasting centre and other key buildings in the capital on the night of the planned coup. After brief fighting in which ten rebels were killed, the uprising was crushed.

In addition, 29 SA ran important operations in Lebanon, where an extensive surveillance mission culminated in the retaliatory killing of terrorists involved in the September 1981 assassination of the French Ambassador in Beirut, Louis Delamère. Travelling separately through various points in Europe and the Middle East, using false identities, some twenty 29 SA operatives entered Beirut during the latter part of 1981. Divided into a 'basic intelligence team', an 'information-gathering team' and an 'action team', they started working through the vipers' nest of private armies and guerrilla factions in the war-ravaged country, to track down Delamère's killers.

Important clues pointed to a Syrian hand in the assassination, according to a member of the basic intelligence team, 'so we concentrated on finding people, mainly young Arab Muslims, who had been rejected or were dissatisfied with some of the Syrian-backed terrorist organizations operating in Lebanon. With their help we were able to plant penetration agents in some of the groups and manipulate them.' The Beirut headquarters and training camps of Abu Nidal and Ahmed Jibril's Popular Front for the Liberation of Palestine General Command, both known for their close links to Syria, were placed under constant surveillance. One French OP which was spotted on an open field observing a terrorist camp was ambushed with machine-gun fire. Both SA operatives were wounded; 'one of them was hit very seriously, and had to be evacuated from Lebanon'.

The eight-month investigation led to a small town in Lebanon's Bekaa Valley, where 'we bought information from people willing to talk about the terrorist cell which killed Delamère'. It was narrowed down to five individuals connected by family ties to a Syrian-paid assassin called Sadek Mousawi. The 29 SA team was able to pinpoint

Mousawi and another gunman, Muhammad Yacine, as the trigger men. Another three or four individuals had driven vehicles, stored arms and otherwise supported the operation against the French Ambassador.

The information-gathering team, divided in four pairs spread throughout the Bekaa, observed and followed the hourly movements of those individuals marked by the basic intelligence team. 'For two weeks we determined where Mousawi and his men lived, their daily routines, where they went, who they met with and what they did.' The detailed intelligence was passed onto the action team, all composed of officers recruited from the 2 REP, who would conduct the terminal phase of the operation.

As Mousawi, Yacine and another individual associated with the terrorist cell got into a Mercedes Sedan on a morning in June 1982 and drove out of Mousawi's residence, two unmarked cars carrying the five Legionnaires of 29 SA's action team, waited, hidden along the street. As the Mercedes slowed down before a crossroads, one vehicle carrying three men screeched up, blocking its path. Mousawi's driver didn't have time to change gear before two silenced FAMAS sub-machine-guns, fired from very close range, ripped through the windscreen, killing him instantly. Simultaneously, the car's back window disntegrated under a spray of rounds fired from the 29 SA vehicle racing up behind.

The drivers of both assault cars kept their engines running while two of the Legionnaires moved along both sides of Mousawi's Mercedes, emptying their FAMAS through the windows and doors until all three occupants lay hopelessly bleeding from multiple wounds. The rear gunner stood watch, covering against a possible reaction force. Jumping back into both cars, the action team sped off in opposite directions. The shooting had lasted no more than fifteen seconds.

But 29 SA had been a little too hasty. Although Yacine and the third man in the car were killed, Mousawi survived his serious wounds, after being rushed to a hospital in Beirut. Hearing about it, 29 SA proceeded to complete their mission. One week after their hit in the Bekaa, two members of the action team entered the Beirut hospital where Mousawi was being treated, disguised as medical staff. Gaining access to the intensive care unit on the second floor, they crept up to his room with silenced pistols. One man guarded the door while the other walked up to the bed where the bandaged terrorist lay comatose and finished him off with a single bullet between the eyes.

The elimination of Sadek Mousawi was only one battle in the deadly secret war developing between French and Syrian special services. Soon afterwards 29 SA was running an operation against Hassan Attich, another terrroist, who had assassinated a DGSE agent and his Lebanese wife in April 1982. Questioning some of Attich's associates, 29 SA was able to confirm that the hit man was operating under the direct orders of the Syrian intelligence chief in Beirut, Major Nouredine.

'It was a question of fighting the terrorists using their own methods,' explains the top DGSE official. 'Israel's Mossad and the British SAS in Northern Ireland had been doing it for years. But it was still difficult to get the idea accepted by a lot of people in the west. The Americans were particularly unreceptive.'

While 29 SA teams operated covertly in the warrens and bombed-out backstreets of Beirut, the secret war extended into Europe, where a series of terrorist attacks during the early eighties targeted French interests. In Paris, crowded cafés, department stores, railway systems, foreign embassies, commercial offices and Jewish synagogues were shattered by bombs and machine-gun attacks. French companies and diplomatic facilities also became frequent terrorist targets in other European capitals.

Before long, 29 SA counter-terrorist teams working on the case uncovered a network of Syrian diplomatic officials and cut-out agents, supporting the terrorist operations against France. 'Personnel in Syrian embassies would smuggle in weapons through the diplomatic pouch, forge passports and provide money to individuals often posing as students who then passed the material on to terrorist teams entering different European countries from the Middle East. The cut-outs would also arrange safe houses and escape routes for the hit men. Libyans and Iraqis were involved in some of the operations but the main support came from Syria.'

The head of DGSE, Pierre Marion, who had pushed for the expansion of the counter-terrorist subdivision, confronted French President François Mitterand with the evidence. 'We have two choices,' he told the President during a personal meeting in the Elysée Palace. 'Either we blow up the Syrian intelligence headquarters in Beirut and eliminate its support elements in Europe or we have to deal directly with Rafaat Assad.' The brother of Syria's President

Hafez Assad and the country's *de facto* Vice-President, Rafaat Assad was known to be the mastermind behind the country's international terrorist network.

As chief of Syria's secret police, the Mukhbarat, Rafaat Assad had once crushed a fundamentalist Muslim insurrection in the northern city of Homm by ordering tanks and artillery to surround the town and bomb it into submission. In flattening Homm, he killed some 10,000 people in a week. He was among the first backers of a Palestinian schoolteacher from Jerusalem called Sabri Al Banna who became an enemy of Yasser Arafat in the PLO and under the *nomme de guerre* of Abu Nidal started the breakaway extremist faction, Fatah Revolutionary Council, responsible for some of the worst terrorist atrocities of the 1980s. A French-educated physician, Rafaat appointed one of his intelligence officers by the name of Ahmed Jibril to create the Popular Front for the Liberation of Palestine General Command (PFLP-GC), a Syrian front group dedicated to supporting the rejectionist line inside the Palestinian movement and organizing hit jobs for the Assad regime.

Rafaat Assad also used his intelligence network to run extensive criminal enterprises ranging from international narcotics trafficking to the marketing of stolen luxury cars. The 29 SA intelligence-gathering operation in Lebanon had updated DGSE's files on his hashish and poppy-growing activity in the Bekaa Valley. The produce was taken in Syrian military trucks to Beirut, where heroin was refined at facilities run by Assad's intelligence operatives and shipped to Europe.

After looking around pensively for a few minutes, Mitterand instructed Marion to have a word with Rafaat Assad. The Syrian often came to France to see an oculist for treatment of his sensitive left eye and agreed to meet the DGSE chief during one of his trips. Fearing an attempt to arrest him, Assad insisted that Marion should come alone and without bodyguards or weapons. The rendezvous was to take place at a château some fifty kilometres outside Paris belonging to Prince Abdallah of the Saudi royal family.

To show that he meant business, Marion instructed his 29 SA team watching a key member of Assad's terrorist network operating under diplomatic cover at the Syrian Embassy in Madrid, to fire on him 'without wounding him'. As Syria's 'cultural attaché' to Spain, Hassan Dayoub walked out of his residence towards his chauffeur-driven Mercedes on the morning of 17 April 1982, he was stunned by the crack of a rifle shot as a single 7.62mm high-velocity round shot past his head, piercing the concrete wall behind him.

A few days later, when the tough but refined head of DGSE drove past the gates of the château he could see dozens of Assad's body-guards posted with sub-machine-guns all over the property. The former French Air Force pilot and executive with Air France had taken the precaution to surround Prince Abdallah's grounds with a 29 SA reaction force. Assad received Marion inside the luxurious mansion, speaking to him in flawless French and both men retired to the study to talk business.

During their two-hour meeting, Assad kept denying any involve-ment in terrorism, complaining that he was being persecuted. Marion finally told him that he knew the identities of the members of his network. 'What happened in Madrid the other morning is just a warning,' he told Assad, 'of what will happen to you and every one of your people on our list.'

After staring poker-faced at Marion for a while, Assad began offering gifts of gold-inlaid damask desks and Arabian horses. Marion abruptly turned him down, stopping the attempt at brib-ery. After a brief monologue in Arabic, the translator sitting between them told Marion that Assad would agree to cease all terrorist activity inside France. 'Syria stopped everything, certainly inside France,' says a top DGSE official, 'at least for the next two years.'

Following the truck bombings which destroyed US Marine and French 2 REP barracks, 29 SA was able to obtain intelligence fairly quickly concerning the Islamic Jihad organization, which had carried out the attacks. Advised by Iranian revolutionary guards and elements of Syrian intelligence, Islamic Jihad was drawing recruits from the most fanatical factions of Lebanon's nationalist Muslim Amal movement, which 29 SA had managed to penetrate. The DGSE presented their intelligence briefing to a meeting of the Centre Opérationnel des Armées, and decided on a retaliatory action for the Beirut truck bombings. DGSE recommended using a covert-action team to conduct individual hits against Islamic Jihad leaders but the suggestion was rejected in favor of an air strike against the group's headquarters.

French Navy jet fighters flying off the aircraft carrier *Foch* bombed a farmhouse in the Bekaa Valley used by the executive committee of Islamic Jihad, but the building was empty at the time and innocent civilians living nearby were killed instead. 'It was a regular military operation and there was advance warning. Too many people knew about it and it failed,' says the top DGSE official. One of the agents which 29 SA had recruited in Amal was found out and killed as a result. 'It was difficult to make our high command understand that you cannot fight terrorism using conventional methods. It's like trying to kill a flee with a bulldozer.'

SEALs were brought back into Lebanon under Lieutenant Commander Tim Holden to advise the Marines in special counter-terrorist weapons use and tactics when they were redeployed in smaller units after the bombings. The GIGN rushed in specialized sniper teams with Philippe Legorjus. But despite a lot of rhetoric, naval bombardments and retaliatory air strikes which led to one US Navy fighter being shot down, the Reagan administration neglected to launch special operations targeting the terrorist cell responsible for bombing the Marine barracks. With some French help, a special Pentagon team known as Intelligence Support Activity set up a surveillance and information-gathering operation in Beirut which pinpointed a certain family connected with the truck bombing. However, the USA took no direct or, as far as is known, indirect action against them.

In a series of subsequent kidnappings over the next few years, Islamic extremists abducted nearly two dozen Westerners, mostly British and Americans. They included the CIA station chief in Beirut, William Buckley, who was tortured and killed; a US Marine officer serving with the UN, who was also found hung from a ceiling and the British hostage negotiator Terry Waite, who was released after several years' captivity. Despite the continuing French presence in Lebanon not a single French citizen was targeted in the wave of kidnappings, as a result of the continued activity of 29 SA.

Covert US special forces teams dispatched repeatedly to locate the hostages determined that most of them were being held in an underground prison complex controlled by Islamic Jihad in the Bekaa Valley. A member of the Amal movement closely connected with the family of its leader, Nabih Berri – who at one point brokered the release of the Spanish Ambassador when he was briefly held by Islamic militants – told this author that he personally saw several of the American hostages in their cells in the Bekaa and informed American intelligence agents.

'There were some efforts to locate and rescue the hostages in Lebanon,' says a top official at the Pentagon's Office of the Assistant

Secretary of Defense for Special Operations and Low Intensity Conflict. 'Special Operations were seriously considered and got to the planning stages.' USAF AC-130 Specter gunships were flown to NATO bases in the Mediterranean, along with C-130 Talon tankers and MH-60 Blackhawk helicopters in preparation for a raid into Lebanon. According to SOS pilots who landed some of the aircraft at the Aviano airbase in Italy, 'We were briefed on flying conditions in Lebanon including hostile radar capabilities.' The operation was going to be conducted with Delta Force. But it was called off. The Reagan administration was instead persuaded to negotiate with Syria and Iran to have the hostages released. This led to the 'Irangate' scandal in which Colonel Oliver North and other officers were forced to resign from the National Security Council.

In December 1983, during the multinational peacekeeping effort in Lebanon, it fell to the French Navy's Nageurs de Combat to rescue Yasser Arafat when he was besieged by Syrian-backed factions of his movement. Arafat and members of his Al Fatah faction had been forced to flee Beirut following an Israeli incursion into Lebanon which had driven the main Palestinian concentrations out of the Lebanese capital. The PLO leader, accompanied by a few hundred of his most loyal followers, had moved to the port of Al Minah to await their evacuation, which was being arranged and protected by the French Navy.

However, soon after setting up his small enclave along the coast, Arafat became surrounded by an army of pro-Syrian Palestinian guerrilla factions waging an internecine war against him. Their objective, supported by the Syrian government, was to assume total control over the Palestinian movement and eliminate conciliatory pro-Western elements grouped around Arafat. Supported by Syrian artillery, they attacked Arafat's positions as French Navy vessels came over the coast and a chartered cargo ship docked in Al Minah to pull Arafat and his people out.

The Israeli Navy, which was observing the situation closely from fast gunboats patrolling off Al Minah, decided that this was the perfect opportunity to let the PLO destroy itself. As night fell, small Israeli patrol boats quietly approached the entrance to Al Minah harbour and dumped about a dozen mines, effectively blocking the channel to prevent Arafat's ship from leaving. If the evacuation vessel was sunk it would naturally be blamed on the Syrians, who would rush into Al Minah to finish off any survivors.

The Israeli mine-laying operation, however, was being secretly watched by an observation team of the French Navy's Commando Huber who had landed some nights previously, setting up hidden positions along entrances to the harbour area. With high-powered infrared night vision scopes, they were able to determine the exact areas where the mines were being laid and what type they were, immediately communicating the information by radio to the French fleet. Twelve more Nageurs de Combat were dropped from a helicopter the next morning, splashing into the waters off Al Minah around the approaches to the harbour in their scuba-diving gear, carrying satchels holding explosives. Diving to the bottom of the channel, they set their underwater demolitions, blowing up the Israeli mines, and Arafat's ship was escorted safely out of Lebanon to sanctuary in Tunisia, from where he negotiated the eventual peace treaty with Israel.

12

THE ELEVENTH HOUR

'All Schwartzkopf was interested in,' according to an SAS officer in the Gulf, 'was tanks, tanks and more tanks.' General Sir Peter de la Billière, the former SAS commander who was called back from retirement to head the British expeditionary force agreed that 'our technological capability on the battlefield and in the air seemed to be so overwhelming as to leave no gap which special forces could usefully fill.' When commanders from SEAL Team Six, Delta Force and the SAS came to sell Schwartzkopf on the idea of very deep-penetration missions behind Iraqi lines they were all told that their services were not required. He did not envisage any major role for special forces in his plan to liberate Kuwait.

'Schwartzkopf had always looked upon special forces as something unmilitary,' according to Colonel Bob Mountel, a former commander of the US Army 5th Special Forces Group, who recalls his arguments with 'The Bear' when they both attended the US General Staff college together. General Schwartzkopf's prejudices about special forces had been reinforced by his experience as one of the planners of the 1983 US invasion of Grenada, in which America's newly revamped special forces failed to perform very impressively. He had seen Delta Force being decimated in their helicopters by Cuban anti-aircraft fire, SEALs drowning on their way to a critical target and Marines being ordered to divert from one of their objectives to rescue Team Six, which was surrounded at the British Governor General's mansion.

Apparently Schwartzkopf was specially adverse to having any SEALs 'in theatre' and when he learned that elements of the California-based SEAL Team One were on board the American Navy's task force steaming into the Gulf he screamed out orders to the effect that none of them be allowed out of the water. Perhaps he didn't know it but USAF SOS Pave Low helicopters had already inserted SEALs along Saudi Arabia's desert border with Kuwait, where the naval teams had operated two hidden front-line OPs since the early days of August, when ten Iraqi divisions were massing for what seemed like a push south. Elements of the Army's 5th Special Forces Group (5SFG) were not far behind, digging into positions further inland along the front in advance of lightly armed US Marines and paratroopers. 'The SEALs were looking for potential Iraqi choke points to conduct operations behind the lines against logistics,' says Colonel Glynn Hale of the 82nd Airborne Division. 'Water supplies were critical and by the time we arrived these were already identified and targeted for sabotage attacks.'

American special forces were also soon busy 'correcting the weaknesses' in the Saudis and other Arab armies joining the US-led coalition. The Arabs needed to be trained in such modern battlefield tactics as Forward Air Control, 'about which they had no concept', according to Lieutenant Mike Riley of the SEALs. 'The atmosphere at Saudi military installations near the front line was like a summer camp. There was no security set-up. Despite the Iraqi Army's proven capability in chemical and bacteriological warfare, they had no equipment or training in this area. If the Iraqis had pushed into Saudi Arabia, it would have been Operation Speed Bump.'

The Arab-language skills of the 5SFG, which was specifically assigned to operations in the Middle East, proved crucial to the training and liaison work required with the coalition forces and Schwartzkopf soon consented to the setting up of a special forces headquarters or Joint Special Operations Task Force (JSOTF) in Dhahran under Army Green Beret Colonel Jesse Johnson. By late October the deputy commander of the 5SFG's two battalions, Lieutenant Colonel Boyd Parsons, was gathered with Egyptian commando officers in a tent at a remote camp along Saudi Arabia's flat desert border with Iraq. The Green Berets were proposing a training exercise, sketching out a plan for the helicopter insertion of a commando team, a nuclear, bacteriological and chemical (NBC) warfare drill, a mock ambush using Milan rockets and RPGs and the team's extraction by helicopter.

During the NBC exercise, supervised by 5SFG officers, smoke grenades were dropped in front of Egyptian commandos advancing through the desert to simulate a bombardment with chemical weapons. The men immediately got down in a defensive perimeter, pulling on gas masks, rubberized boots, gloves and protective overalls as their patrol leader raised his headquarters over the field radio to report the attack and request to move around the contaminated area. His orders were, instead, to survey the zone to determine the degree of contamination and type of chemical agent. The ten men moved slowly forward in single file, crossing the smoke-covered patch of Mars-like moonscape as two of them measured the air with special chronometers. They marked the danger zone and the lieutenant reported his findings back by radio. Once past the contamination zone, the commandos stripped off their NBC kit, buried the contaminated gear in the ground and moved on.

Although the Egyptians' commando training had recently been influenced by American special forces, the Soviet disruptive pattern of their camouflage uniforms contrasted with the 'chocolate chip' desert wear of their American advisers. Their standard assault weapons, the AKMS 7.62mm rifle with collapsible stock and the Tokagypt 9mm pistol, are also of Soviet design, as is the SA-7 anti-aircraft missile carried by jeep-mounted patrols. The similarity of the Egyptian armaments to those of the largely Soviet-equipped Iraqi Army 'could prove useful in operations behind the lines', said Parsons.

The Green Berets at the commando base just north of the frontier town of Hafr Al Batin squatted cross-legged around a low, T-shaped table to feast on a banquet of roast baby lamb served on giant platters. The occasion was a visit by the Commander of Egyptian Forces in Saudi Arabia, General Muhammad Ali Belal. A commando himself, the black-skinned Belal had led the daring crossing of the Suez Canal to reconquer the Sinai Desert in Egypt's 1973 'Yom Kippur War' against Israel. He refused to refer to the Iraqis as 'enemies', preferring to see them as 'Arab brothers who have committed an aggression'. It was particularly hard for Belal because he had also served as an adviser to the Iraqi Army during Saddam Hussein's ten-year war against Iran.

American special forces teams were being assigned at battalion level with every Arab army deployed in Saudi Arabia, explained Parsons as he tore a chunk of meat off the skewered lamb lying in front of him with upturned feet. We ate with our fingers, bedouin style. He described his job as 'trying to make sure they're all pulling the oars together'. The only ones resisting the introduction of Green Berets down to unit level were the Syrians, whose army was considered among the most effective in the Arab coalition.

While the Green Berets upgraded Arab capabilities in such fields as combat air control and communications, they also gathered valuable intelligence about their allies and about the Iraqis. Egyptian commando patrols were intercepting the radio communications of Iraqi front-line positions and although the Allied offensive was still three months away, one Egyptian officer told me that Saddam's army was already breaking down. 'The units in Kuwait do not have enough water, enough food, enough fuel. Their morale is very bad.' Belal believed that the Iraqis would not last for more than a few days if it came to a fight and thought that Saddam would have enough sense to pull out of Kuwait before that happened.

Indeed, by mid-November it seemed that diplomatic efforts to bring about a negotiated settlement were bearing fruit. As a good-will gesture, Iraq's brutal dictator had even dropped his 'human shield' of thousands of Western hostages he had trapped in Iraq and Kuwait, allowing the civilian men, women and children to return home to their respective countries. This relieved the SAS of the only role which General de la Billière had developed for them – planning a nearly impossible rescue operation. 'No rescue operation could ever have been a complete success,' he writes. 'With the help of the Foreign Office and the Ministry of Defence, the SAS worked hard to build a picture of where the prisoners were being held. To have had any chance of success, they would have had to go in by parachute or helicopter and lift the hostages out to safe rendezvous in the desert – as the Americans tried to do in their attempt to rescue their people from the Embassy in Tehran in 1980. But at no stage did I feel that we could have recovered more than half the [2000] Britons, if that.'

Two squadrons of the SAS and a headquarters element which had been flown up to the United Arab Emirates for 'desert training' were sent packing back to Hereford and the men spent Christmas with their families. Maureen Denbury, the mother of an SAS corporal, recalls walking out of the kitchen of the family-owned pub in Wales to serve the usual lunch-time crowd one mid-December day to find her son David sitting at the bar with a deep suntan. He had ordered the plate of chips and curry sauce which she was carrying. Mrs Denbury hadn't seen him since he had left for the Middle East five weeks after the Iraqi invasion. 'Memories of that Christmas will stay with us for ever,' she says. 'It was the last time that we saw David alive.'

As diplomatic negotiations started breaking down and the 15 January deadline set by President Bush for Iraq's withdrawal from Kuwait drew near, Schwartzkopf began considering the use of special forces to achieve some of his objectives. The most immediate of these was clearing an air corridor through Iraq's early-warning radar system through which the first wave of Allied aircraft could fly to reach Baghdad undetected. This would have the double purpose of catching all enemy targets by surprise and minimizing casualties among Allied pilots during the initial stages of the air war – an important political priority.

A clandestine raid by the 5SFG was planned against two early-warning radar sites operating along Iraq's western border with Saudi Arabia. A small American special forces team with possible assistance from Egyptian commandos would cross the border by land seventy-two hours ahead of H hour and blow them up. But the precise timing required for the raid meant that the 5SFG had to be one hundred per cent certain of being on target no sooner and no later than twenty minutes before the Allied air strikes were launched. There was little margin to allow for the possibility of their being delayed by map-reading errors while navigating through the flat, featureless desert. Going into Iraq too far ahead of time, on the other hand, increased their chances of a compromise.

The special forces group needed to be equipped with Magellan Global Positioning Systems (GPS), to give them an accurate satellite reading of their distance and bearings from a target down to 100 feet. But the hand-held variety of GPS was scarce at the start of the Gulf War and because of limitations in satellite coverage, they worked only four hours a day.

Becoming impatient with all the risks and deficiencies in the plan, Schwartzkopf was about to cancel the mission when the head of JSOTF, Colonel Jesse Johnson, suggested using SOS Pave Low helicopters with built-in GPS. They could fly from the special forces forward base of Al-Jouf into Iraqi airspace below radar cover and have the certainty of locating the two radars and their command centre within a tightly programmed schedule. The only problem was that a noisy helicopter landing of the sabotage team could raise an alarm at the radar installation manned by 400 personnel and trigger a call to central headquarters in Baghdad before the communications lines were destroyed. If alerted about the attack, the Iraqis could cover the estimated seven to eight-mile blindspot which would be opened for aircraft flying at 1000 feet, by reconfiguring other radar systems, defeating the whole purpose of the raid.

At first Johnson tried to convince Schwartzkopf that two or three Pave Lows could destroy the radars with just their side-firing .50-cal machine-guns without the need for a ground assault. But the helicopter pilots tasked with the mission pointed out that, 'we can go in there and put the things out of commission but can't guarantee that we will do it without alerting Baghdad. For that you need real explosive power and highly accurate stand-off systems to knock out radio antennae and telephone lines which the lightly armed Pave Low does not have.' Then Captain Randy O'Boyle, the Air Force Special Operations Command's liaison officer with Schwartzkopf's staff, came up with the idea of having Pave Lows guide in a column of the Apache attack helicopters armed with laser-guided Hellfire missiles, 70mm rockets and 30mm Gatling cannons, which could knock out the radars and all their ancillary equipment with the necessary speed and lightning force.

Navigating on their GPS and terrain-following/terrain-avoidance radar, two Pave Lows practised flying as pathfinders for eight Apaches detached from the 106th Aviation Battalion of the 101st Airborne Division. Hovering below 100 feet at night, they would guide the attack helicopters to within twelve kilometres of a target, from where it would be acquired on the Apaches' infrared cameras so that they could release their devastating fire-power with pinpoint accuracy. After two rehearsals at a Saudi practice range where a total of nine laser-guided Hellfire missiles (one million dollars' worth) were fired, the Pave Low and Apache teams felt confident that they could work together on a joint mission.

'All right, brief me,' said Schwartzkopf with a prove-to-me-that-you-can-do-it look, as he sat down on the couch in his office at the Saudi Air Ministry. Kneeling before their portly supreme commander, two Air Force special operations officers and Colonel Johnson, spread out a three-fifths map on the floor and gave a detailed presentation of their plan. At the end of the twenty-minute briefing, Schwartzkopf just grunted, 'Uh-uh', stood up, went to his desk and asked Johnson to stay behind. During their private meeting, which went on for half an hour, Schwartzkopf said he had to have a one hundred per cent guarantee of success from JSOTF. Johnson gave it.

At 21.30 on the night of 16 January 1991, the two Pave Low

helicopter crews were given their final briefing. They had completed their detailed planning a week before, when JSOTF cleared them for the target location and other full information on the mission, and a realistic dress rehearsal with the Apaches was run on 11 January. All the helicopters were positioned for take-off on the runway of Al-Jouf at 00.55. H hour for the coalition's air offensive was set for 03.00, which meant that the helicopters were scheduled to hit the radar sites at 02.38. Air operations during the Gulf War were timed to the minute.

Oh God, please don't let me fuck up, thought the lead pilot, Captain Mike Kingsley, as he strapped himself into his seat and began the flight checks. He felt a moment of panic when the electrical power in his Pave Low went off line while he positioned the airborne alignments of his radar and GPS with the satellites. But it came back on in five seconds.

Lieutenant Colonel Richard Comer, who had personally briefed Schwartzkopf and was riding as co-pilot in the lead Pave Low, remembers going through all the imponderable questions in his mind. Would they be waiting for us? Had there been security leaks? Satellite reconnaissance of the radar sites indicated that there were only machine-gun positions protecting them. But could there be heat-seeking missiles hidden nearby, perhaps in that bedouin camp seen in the intelligence photographs. There was an Iraqi airbase only ten minutes' flying distance away at jet-fighter speed. Would they come after us? 'Only if they were waiting for us because within five minutes of our time on target, the skies would be full of F-111s, F-16s, F-15s and all the other aircraft of the mightiest air armada ever assembled.'

Cruise missiles fired long distance by B-52 bombers and Toma-hawks blasting off US Navy ships were already on their way to targets deep inside Iraq as the two SOS Pave Lows became the first manned Allied aircraft to cross into Iraqi airspace. The pilots flew the helicopters some eighty feet above the ground, or 100 feet below standard radar cover, scrutinizing their forward-looking infrared sensors to spot their targets. They were a dozen kilometres inside Iraq when the pilots detected the heat signature of the enemy radar sites positioned at the top of a 200-foot knoll rising over the desert well above their altitude. Within ten kilometres of the targets, the Pave Low pilots flashed an infrared beam which was picked up on the control panels of the four Apaches following some two kilometres behind. The Pave Low tail-gunners then dropped a bundle of luminescent 'kimsticks' to mark the spot for the Apaches to release their Hellfires.

The Pave Lows banked left and right, flying some three kilometres before going into orbit and seeing the entire desert lit up from the impact of eight Hellfires obliterating the radar complex. 'For anyone on the ground, it would have been like the world coming apart all around them,' says Comer. 'They wouldn't even know what hit them.' With their lights off, the Apaches were invisible in the night sky at their distance from the targets.

Immediately after releasing their Hellfires the Apaches switched on outside lights, to avoid possible collisions, as they gained speed and hovered uphill to strafe the radar complex with their 70mm rockets and 30mm cannons. After slaughtering all the personnel and making sure that no single radio antennae or telephone line remained standing, they flew several passes over the burning wreckage. It was seven minutes after the first Hellfire volley that Colonel Johnson received confirmation from Comer, thorough a direct radio link between the Pave Low and the war room in Riyadh, that the mission was accomplished.

But no communication could be established with the second Pave Low because its crew was too busy with an evasive manoeuvre to avoid being hit by two Soviet-made SA-7s. The heat-seeking, shoulder-fired missiles had suddenly streaked up towards them from a concealed Iraqi position in the bedouin camp a few kilometres from the radar site. Infrared energy from the Pave Low's package of electronic counter-measures succeeded in deflecting the missiles and its pilot came back on the radio some minutes later.

Crossing back into Saudi Arabia, the helicopter pilots could hear the gathering drone from the air fleet flying some 900 feet above them, heading into the air corridor they had just cleared. Lieutenant Colonel J.V.O. Weaver, 'The Mayor' of the special forces base at Al-Jouf, calls it 'the most flawless attack in history'. He stood on the runway at first light as the Pave Lows, Apaches and two MH-60 Blackhawks which had tailed behind for possible emergency rescue purposes, hovered safely back on to the airfield. 'The guys came back starry-eyed,' he remembers one of the youngest exclaiming. 'God, you should have seen what we just did – fucking awesome.'

In the 3000 Allied air sorties flown over Iraq and Kuwait on the first night of the war, only two warplanes got shot down, a Navy F-18

Tomcat and an Italian jet fighter from the Air Squadron which Italy had sent as its sole contribution to Desert Storm. Not being American or British, the captured Italian pilot was not even of much propaganda value to Saddam Hussein.

While the Pave Lows and Apaches had been revving up their engines in Al-Jouf, Lieutenant General Sir Peter de la Billière was being ushered into a plush office suite at the Air Ministry in Riyadh for a private meeting with Prince Khalid, overall commander of the Arab forces. The next in line to the Saudi throne looked visibly worried and after some pleasant words of appreciation of what Britain was doing, he began imploring de la Billière to pass on to his government the overriding need to keep Israel out of the war. 'If the Israelis come in, it will be exceedingly dangerous. It will split the alliance. It will be a disaster,' said Khalid.

Saddam Hussein knew perfectly well that his army would not prevail militarily against the overwhelming fire-power ranged against him. But he believed that Iraq could steal a political victory by what amounted to a strategy of terrorism, something with which the Iraqi dictator was intimately familiar. His medium-range Scud missile force would strike Israel's main cities from launching sites in western Iraq, provoking an Israeli retaliation or perhaps invasion of the area which would realize Prince Khalid's worst fears. Schwartzkopf's meticulous computer-controlled war plans would be short-circuited by a man who had begun his meteoric rise as a lone assassin for Iraq's radical Baath Party and who knew the importance of the unforeseen human element in any armed conflict since the day that he had killed Iraq's last monarch.

De la Billière had been patiently working on Schwartzkopf to gain acceptance for the idea of launching some long-range penetration missions deep into Iraq. The former SAS commander, along with some high-level officials at the Pentagon, was becoming convinced that given Saddam's ruthlessness and unpredictability, on-ground surveillance and direct-action missions were needed on strategic targets such as his Scud sites and battery emplacements for chemical and bacteriological weapons. South African-manufactured G-5 howitzers used by the Iraqi Army to deliver chemical and bacteriological shells were hardly distinguishable from conventional artillery by air reconnaissance. The SAS had got hold of a G-5 to learn its operation and distinguishing components.

Schwartzkopf considered the inaccurate and clumsy Scud as a

militarily insignificant weapon, and was convinced that its threat
could be neutralized with air strikes. His answer to the scenarios
involving chemical attacks was to threaten Saddam with nuclear
retaliation if he resorted to the use of unconventional weapons.
Oblique warnings to this effect were made by various Western
political leaders. But as the 15 January deadline drew near, it became
clear that Western threats were failing to impress Saddam, and
Schwartzkopf finally consented to bringing in the SAS to Saudi
Arabia, as a strategic reserve for possible operations to support an
eventual land offensive.

On 29 December, Corporal David Denbury bid his final farewells to
his family and kissed his girlfriend Ceri goodbye before mounting his
motorcycle and riding away from the Star Inn for the last time.
Arriving back in Hereford he rejoined A Squadron, which during the
next two weeks was boarded on to trucks along with B and D
Squadrons to be taken on to waiting VC-10 transports planes at the
base of RAF Brize Norton. From there the SAS was flown to Saudi
Arabia on several airlifts. Sergeant Andy McNab of B Squadron
describes the atmosphere at the Riyadh airport hangar stinking of
diesel fumes where he was billeted along with forty other men on
arrival as 'jovial and lively. The regiment hadn't been massed like this
since the Second World War. It was wonderful that so many of us
were together. So often we work in small groups of a covert nature,
but here was the chance to be out in the open in large numbers.
Classic behind the lines soldiering was what David Stirling had set the
Regiment up for in the first place and now fifty years later, we were
back where we started.' McNab adds, 'But we didn't have a clue yet
about what we'd have to do.'
     In the days leading up to the Allied air offensive, while American
special forces helicopters were gathering at Al-Jouf to launch their
opening salvo, the SAS remained in limbo. No specific missions had
yet been assigned to the Regiment. In Riyadh, the SAS commander,
Colonel David Holmes, would regularly drop in to see de la Billière at
his office in the Air Ministry to get the latest news. The greying
general would usually stroll out with the quiet, unassuming Holmes.
They discussed the possible behind-the-lines missions being considered by Schwartzkopf, such as creating a diversion for the Iraqis
along the western flank of the British Armoured Division. But the
plan-driven American general would still not give the go-ahead for

the SAS to enter Iraq, remaining reluctant about committing any ground troops too far ahead of the land offensive.

'But when the first Iraqi Scuds hit Tel Aviv, Riyadh and Dhahran simultaneously on the second night of the war,' says an SAS officer who served in the Gulf, 'everybody was knocked off their seats.' Twelve Scuds had been unleashed on Israel alone. Decontamination teams with Geiger counters rode through residential suburbs hit around Tel Aviv, measuring the air for possible chemical or bacteriological agents. The threatening possibility that the children of Hitler's Holocaust were having a 'take-out' gas chamber delivered to them by a new exterminator of their race was a prospect which struck a very raw nerve with Israel.

Israeli Defence Minister Moshe Arens was on the phone to US Defense Secretary Dick Cheney asking him for the recognition codes which would allow a squadron of Israeli F-16s warming up their engines at an airbase outside Tel Aviv to move through Allied war planes to attack Iraq without being shot down. Israeli airborne units also went on alert for possible commando operations in western Iraq. The situation was desperate and Schwartzkopf was seriously embarrassed because his headquarters had earlier claimed that Iraq's Scud capability had been neutralized in the first wave of Allied air strikes. The after-action reports which had fed such statements had not been deliberately misleading. The pilots just hadn't realized that the various Scud sites identified by satellite photography which they hit, were, in fact, very convincing decoys manufactured in East Germany and supplied with the Scud package bought from the Soviet Union.

At his first press conference since start of the war, on 19 January Schwartzkopf admitted that looking for individual Scud launchers in the vast desert was like 'finding a needle in the haystack'. Conceding that events had somewhat clouded his intricate planning, he said, 'The fog of war is present, the picture is not perfect.'

Cheney as well as the Chairman of the Joint Chiefs of Staff, General Colin Powell, had already been telling Schwartzkopf over the red phone connecting his war room with the Pentagon to use all available means to go after the Scuds, including the deployment of special forces on the ground – ASAP. 'Sending the SAS and Delta into western Iraq took prodding from the highest levels in the Defense Department,' according to a top official of the Pentagon's Office for Special Operations and Low Intensity Conflict. 'Schwartzkopf had to be told by Washington that neutralizing the Scud threat was no longer a strictly military matter but had become a political priority.' A top SAS officer in the Gulf maintains that 'without the Scuds, the SAS role in the war would have been non-existent'.

On the morning of 19 January, all three SAS squadrons lifted off from Riyadh in C-130s for their final 900-mile flight north-west to Al-Jouf. 'The Hercules aircraft transporting the men had to pick their way across the flight paths of war planes heading for Iraq,' according to de la Billière, 'and they flew at low level under American control every inch of the way through windows designated by the US Tactical Air Control Center.' Delta Force, meanwhile, was flying out of Fort Bragg, North Carolina to link up with the SAS for the joint operations being hastily planned against the Scud sites, whose communications, command and control and support network were scattered over several thousand square kilometres of desert along Iraq's border with Jordan and Syria.

'Special forces should have gone in a lot sooner,' believes Colonel Bob Mountel of 5SFG. 'If they had been systematically employed collecting information and intelligence behind the lines prior to hostilities, they could have pinpointed the Scuds, their communications lines and guidance systems well in advance and had a deep reconnaissance picture of western Iraq already prepared.' But when Corporal David Denbury, lead scout for A Squadron, kick-started his motorcycle, riding off into the overcast horizons of the cold, flat desert as one of the first special operatives to enter Iraq, his troop commanders and sergeant majors lumbering behind in heavily loaded jeeps were directing the operation virtually blind.

According to a high-ranking SAS officer in the Gulf, '1/100,000 relief maps were the most detailed guides we had on our vast zone of operations'. These were assigned to four half-squadron fighting columns which started driving across the border on 20 January in jeeps mounted with Milan rockets, .50-cal machine-guns and 40mm grenade launchers. Loaded down with over a ton of war material, each vehicle had enough capacity to carry three men. 'We had virtually no hard intelligence on Iraq's rocket force,' says the SAS source in Al-Jouf. 'No names or profiles of the commanders or description of the units and their organization. We had only vague grid references about suspected Scud locations in areas designated H1, H2 and H3. The needle in the haystack feeling stayed with us through to the end.'

13

DENIED TERRITORY

A number of small special forces teams were inserted by helicopter deep into Iraq for silent reconnaissance and sabotage operations. Among the main targets were the strategic underground secure communications lines running east and west of Baghdad. These land-lines, made of fibre-optic cable, were invulnerable to air attack or electronic eavesdropping and after other command and control systems had been severed, could still allow Saddam to retain control over his Scud batteries in western Iraq and elite Republican Guard divisions positioned with chemical weapons just north of Kuwait.

Sergeant Andy McNab's eight-man patrol, code-named Bravo Two Zero, which was tasked with destroying the western land-lines, had decided among themselves to be choppered in and left to operate on foot 180 kilometres inside Iraq because it would take too long to drive there. With just a rough approximation of the targeted land-line's location – it was known to run along an Iraqi main supply route (MSR) which they would also be observing – Bravo Two Zero prepared an elaborate sabotage plan. 'We worked out that, depending on the ground, we'd do an array of four, five or six cuts along the cable and each one of them would be timed to detonate at different times over a period of days. We'd lay all the charges in one night and have one going off, say, in the early evening the next day. That would give one whole night when, at best, it was incapable of being repaired and they'd come, probably at first light, to fix it. They'd eventually find out where the cuts had been made and send a team down to repair them. It made sense for us to try to include these people in the damage, thereby reducing the Iraqis' capability to carry out other repairs. Mark came up with the idea of putting down Elsie mines, which are small anti-personnel mines which work on pressure. When you step on them they explode.'

Cutting the eastern land-line was tasked to the Royal Marines' Special Boat Squadron operating with the British naval task force in the waters of the Gulf. The SBS team, assisted by a US Special Forces Combat Air Control officer, flew into Iraq in two RAF Chinooks at an altitude of fifty feet, passing over an Iraqi military convoy on the way. 'The lead vehicle had his lights on and the seven trucks following had theirs switched off. They could hear us but not see us as our helicopters were blacked out.' They landed at a point thirty-five kilometres further north along the Baghdad–Basra highway some fifty kilometres outside the Iraqi capital. It was still dark on the morning of 23 January. 'We could see our fighters going in and egressing from Baghdad. Red tracers from AAA fire shot out into the sky from the city's defences.'

The Chinooks remained on the ground with their engines running but rotors disengaged to cut down the noise, waiting for the sabotage team to complete their task and lift them out immediately. The SBS men quickly ran to a point some fifty metres closer to the highway and spread out in pairs, using electronic detectors to search for the hidden cable. Meanwhile the American special forces sergeant set up his three radios on the ground, tuning into the frequencies of fighters flying overhead to be prepared to call in emergency air cover.

Locating the fibre-optic cable within half an hour, the Marines dug up the ground at several points using pneumatic mini-diggers. Suddenly tracer rounds from an Iraqi anti-aircraft position some five kilometres away came right overhead, 'burning up right on top of us'. Sergeant Jones raised an F-15's call sign, but the anti-aircraft fire died down and no air strike needed to be called in.

The SBS finished their job in the next hour. Exposing the cable, they cut off a length of the meshed fibre-optic wires to take back for analysis and attached demolitions charges to the other portions they had dug up. Dawn was breaking as the Chinooks extracting the SBS team were safely back over the waters of the Gulf and a series of explosions blew out a mile-long stretch of Saddam Hussein's secret communications network. The land-line's above-ground marker, which the SBS commander took with him, was sent to General Schwartzkopf as a souvenir.

Lightning raids on other sensitive installations were executed by helicopter-borne US special forces teams throughout Iraq. JSOTF had secretly brought in a shock unit of seventy-five Rangers to support direct-action missions. One assault targeted a communications complex on the Jordanian border handling important message traffic between Baghdad and the fence-sitting government of Jordan, which the Riyadh war room had reason to believe was passing advance warning on American aircraft flying in from carriers in the Red Sea.

Minutes after an F-15 surgical strike neutralized the installation's air defences, two Pave Lows escorted by Cobra helicopter gunships landed directly on the installation. As the Cobras hovered over the complex, spitting up chain-gun fire on to the scattered Iraqi troops and gun emplacements, the Pave Lows touched down, their rotor blades stopping barely fifteen feet from the concrete walls of the command bunker. A platoon of Rangers ran out with NVGs strapped over their eyes, carrying M-16s and a few MP-5s. Some fanned out while others blew open the door to the bunker and rushed inside. Surrendering Iraqis were forced to the floor and quickly interrogated as Delta intelligence specialists opened computers, located files and rummaged through the equipment. As soon as documents, tapes, radios and other gadgetry were retrieved to be taken for analysis, all the men piled back on board the Pave Lows and took off into the night.

Things were not quite so easy for the Delta teams on the ground, which at any one time ran up to eight OPs and were soon getting into some heavy eight-hour fire-fights. In one episode reminiscent of Vietnam days, a patrol was lifted out by a Blackhawk in the midst of an intense battle with an Iraqi company encircling their position. The helicopter landed directly between the six besieged Delta men and a force of charging Iraqis. Spraying thousands of rounds, the side-window machine-gunner threw back the enemy long enough for the Delta men to rush inside. In another intense engagement a Delta sergeant 'got shot to shit', being evacuated by helicopter with several serious gunshot wounds, but survived after some intensive surgery.

'At one point I had twenty-three people missing in action,' says an SAS officer in the Gulf. 'There were lots of Custer's last stands out there,' concurs Colonel Parsons of the 5th Special Forces Group. 'Out of a dozen American special forces teams sent into Iraq, most of them found themselves in extreme circumstances. It's a miracle they all came back . . . There were some horrendous fire-fights going on out there.'

Soon after being dropped off 187 miles behind enemy lines, Sergeant Andy McNab and the seven other men of Bravo Two Zero found that the ground was a lot flatter than shown on their intelligence maps and the zone was more populated than their brief had indicated. The leader of another SAS foot patrol being inserted some forty miles to the south of McNab's position found the area so flat as to make concealment impossible. He immediately ordered his men back into the Chinook and returned to Al-Jouf.

But McNab persisted and just a few hours after his patrol established their OP in a shallow cave in a low, rocky escarpment, Iraqi soldiers came into view, setting up an anti-aircraft battery between the OP and the MSR, less than 500 metres away, which the SAS was observing. The patrol hadn't even started searching for the land-lines before a herd of goats belonging to a local farmer stampeded through their position and the soldiers were spotted by a young boy looking for a stray calf. 'We are the SAS, not the SS,' says McNab, explaining his decision not to shoot the child as he ran off towards the Iraqi AA battery.

Eight Green Berets of the 5SFG's Alpha patrol faced similar problems when they were inserted 150 miles inside Iraq to observe enemy movements along Highway 7, running from Baghdad to the southern city of Nizariya along the Euphrates Valley. 'We had gotten a picture of the infiltration area which showed that there was nothing there,' says the team's communications specialist, Sergeant James Weatherford, 'but something must have been wrong because we found ourselves in the middle of a town.' As dawn broke following their night-time helicopter landing, the men realized that buildings close to their OP were not the 'ruins' which had been marked in the satellite photos but a farming village. 'As morning set in, it became apparent the fields around us were heavily travelled,' says Alpha patrol leader Sergeant Richard Balwanz. 'Children were playing, women were gathering wood, many were herding sheep, goats and cattle. No matter what direction one looked there were lots of people moving about.'

Operational detachment Alpha 525 had loaded on to two MH-60 Blackhawk helicopters at 23.14 hours the previous night, taking off from King Khaled Military City. The lights of the vast complex serving as a forward base for the coalition's Arab armies, which includes a palace for the Saudi royal family, soon disappeared below them as they flew out over the surrounding black emptiness of the desert. The helicopters refuelled at a small airfield further north for the long flight into Iraq, lifting off from there just after midnight.

Months ago, during the 5SFG's initial deployment to Saudi Arabia, Balwanz had asked his commander to be given the first mission behind the lines. His 'Sharkmen' – they were all scuba-qualified – trained hard in perfecting hide-site systems, combat air control and practised with the latest sniper weapons, including the McMillan .50-cal high-speed armour-piercing rifle. They wouldn't settle for anything that wasn't the real thing, turning down jobs like foreign internal defence, which they didn't consider good enough. 'We were arrogant,' says Weatherford, a slender twenty-seven-year-old with pierced ears who wouldn't look out of place in a disco and had joined special forces 'because the rest of the army was too slow for me'.

When they were assigned their mission after war broke out, the eight men went into isolation with their helicopter pilots to plan every aspect of the operation down to the finest detail. They mapped out the best insertion and emergency extraction routes to and from their objective. Studying each man's bowel movements, they calculated how many plastic bags and condoms they would need for the mission's expected six-day duration. 'We would shit in the bags and piss in the rubbers. Each bag was filled with a bit of lime juice to kill the odour and make it decompose faster.'

'The planning was left mostly up to us,' says Weatherford, who acquired HF radios, Satcoms and additional PRC 90 emergency radios and spare batteries to spread around the team and 'make sure that everybody could talk to someone'. Each man would be carrying 160-175lb of equipment, including five gallons of water. Once their briefback was accepted by headquarters, the eight men were airlifted to King Khaled Military City, where they waited for a week in an aircraft hangar before getting the final go-ahead.

Soon after crossing the border into Iraq, their two helicopters were recalled to the Saudi airfield of Rafha because some general had not yet given his authorization for the mission. Once the formality was completed the Blackhawks refuelled and lifted off again into Iraq. 'This delay not only placed the team forty-five minutes late on target,' explains Balwanz, 'it caused a loss of satellite coverage for the Ground Positioning Satellite system. The helicopter flight leader informed me that due to the satellite coverage loss, he could not place us directly on the infiltration landing zone.'

Flying at an altitude of ten to twenty feet, one chopper almost hit a sand dune, but the pilot managed to jerk it up just in time to avoid a major accident. Flying over a herd of camels, the men could hear the animals' cries through the noise of the engines. After the standard deception technique of making a false insertion five minutes before reaching the landing zone, the helicopters' back ramps came down and the eight men emerged, moving cautiously, wearing black headscarves over their faces, which were smeared with camouflage grease. Some stood guard while others lugged off their heavy equipment. The patrol's hand-held GPS indicated that they were at a point some 1500 metres north of their objective. 'You feel very isolated when the helo lifts up and you are deep inside enemy territory,' says Balwanz. But Sergeant Terry Harris felt like it 'could be just another training mission. Just different aggressors'.

After making a brief halt and security check in which they cached an HF PRC 104 radio under the ground, they started their night march. 'Walking a mile with all our equipment seemed like an eternity,' comments Balwanz. After about a kilometre the team leader and his second in command, Sergeant Charles Hopkins, armed with silenced Heckler & Koch MP-5 sub-machine-guns, moved forward to conduct the leaders' reconnaissance. The rest stayed in covering positions with their M-16 automatic rifles, two of them fitted with 203 grenade launchers. As is standard for operations behind the lines, all eight Green Berets also carried 9mm pistols tucked in shoulder holsters.

'We found ourselves in flat, level agricultural fields with criss-crossing drainage canal systems. The soil was of packed dirt which was wet with puddles of standing water, dampening our clothes and boots. Visibility was excellent in all directions and illumination was 71 per cent. We conducted all our movement in the ditches where possible to keep down and off the open fields. The canals were anywhere from one foot to six foot in depth and up to four feet wide. It was very quiet and there seemed to be no one in the area. Some dogs were barking but they soon quieted down.'

Balwanz decided that the best place for the hide-sites was along the sides of a large ditch where dirt was piled up on either side and the soil was looser, making for quicker digging. One position was some 250 metres from Highway 7 and the other was further back at about 325 metres. The men worked as fast as possible to dig their two hides.

Due to the delays in getting out of Saudi Arabia, Alpha patrol had only a few hours of darkness left in which to complete the construction and couldn't make the hides as deep as they would have wanted. The soil also turned out to be much harder than they had anticipated.

'We wanted to dig down to mid chest but the ground and time constraints only permitted us to be waist deep before it was light.'

While exercising in the Saudi desert, Alpha had practised setting up hide-sites with plastic, burlap and other material held off the ground by a pole, forming a dome which when covered with dirt looked virtually indistinguishable from a sand dune. But the camouflage architecture did not blend in as well with the grazing fields of the Euphrates Valley. As the morning wore on and the Green Berets found themselves in the midst of cattle herds and farming families going about their morning chores, Weatherford turned his head away from the hole's view port, telling Balwanz, 'I don't think we will make it till dark but if we do we better find another location.'

As the Green Berets observed and recorded enemy traffic along Highway 7, three children were heard playing just a few metres away. The two little girls and a boy between the ages of eight and ten could not be seen from the view port but their voices drew closer while at the same time getting lower as they noticed something strange about the earth. One little girl saw the view port and looked directly inside, staring right into a pair of binoculars and Weatherford's camouflage-painted face. She screamed.

It is described by an SAS commander as 'the no-win question' asked of all candidates during the psychological phase of special forces selection. It could fail the toughest of men by exposing their inability to squarely face up the dilemma which during the Gulf war had to be confronted by at least two special forces patrols in a real combat environment. Killing the child would mean becoming a murderous brute, morally no better than the worst enemy one could be fighting, while not to eliminate the threat of compromise could destroy the mission. Bo Gritz, Vietnam War hero of the 5SFG, when coming across the womenfolk and babies of an ambushed VC column while operating in dense enemy jungle, had said 'the very worst thing for a combatant to do is take the life of another human being who is not suited up for the game. The needless death of those not privileged to be soldiers is no victory for the warrior.'

'Kostrzebski, don't even think about, it's just kids,' whispered Weatherford to the Green Beret next to him reaching for the silenced MP-5. Like his SAS counterpart in Bravo Two Zero, Sergeant Balwanz agreed, saying, 'We are not here to shoot children', and let the little girl run away. The team leader instructed Weatherford to contact headquarters. Talking to their special forces control officer at the JSOTF base thirty-five miles outside Dhahran via Satcom, Balwanz reported, 'Position is compromised, we want emergency exfil.'

'We will see what we can do and get back to you,' came the lieutenant colonel's reply.

In the other hide, Sergeant Terry Harris had turned away momentarily from the view port to log two military trucks he had just seen going down the highway. The vehicles and their time of passing would be reported back at standard communicating time every six hours. Only if they saw a Scud or a lot of heavy artillery should they report it immediately. 'Suddenly I saw Balwanz come running down the ditch, informing us of the compromise and plans for an emergency exfil.' They had to evacuate the hide-sites immediately. Rucksacks were quickly repacked with radios and other equipment as both groups of Sharkmen met up in the ditch. It was chest deep and provided good cover.

As the Green Berets cautiously moved 300-400 metres away from the abandoned positions, wading into waist-deep mud, they saw that the large number of civilians working the fields around them seemed to be entirely undisturbed, oblivious to the patrol's presence. Balwanz decided that the children had failed to raise an alarm and the patrol should remain concealed in the canal for the rest of the day, observing the highway from there. At night, they would move further north and set up new hide-sites. He picked up the Satcom and instructed headquarters to put the emergency exfil on hold.

Cows mooed and sheep grazed peacefully, and two old women gathering crops passed right by the eight men hiding in the wet ditch and carried on as if they had seen nothing unusual. The fields were turning clammy under the mid-morning sun, making the men's damp, muddy uniforms feel steamy. Three hours had passed and Alpha made their scheduled contact with headquarters, informing them that there was no significant movement on Highway 7. At about 10.00 Balwanz noticed a crowd of civilians moving from the north-east towards their position. People were also gathering around the two abandoned hide-sites. An older Arab in the usual white robe with a checked *shemag* wrapped around his head, accompanied by two young boys, drew near. One of the boys saw the Green Berets in the ditch and exclaimed something. The man came over and looked directly at the Green Berets, saying some words in Arabic. The Americans politely greeted him with '*Salama a la Kum*', God's peace be upon you.

The Arab turned his back and walked away towards the village at a rapid pace, herding the two boys and the sheep away from the ditch. The men looked at each other and then at Balwanz. 'There is something about shooting a man in the back which doesn't sit well with Americans,' says Balwanz. 'Just couldn't do it'. The Sharkmen moved further east along the canal but stopped when they noticed a crowd of people ahead looking in their direction and for the first time saw two uniformed Iraqis among them. The Green Berets turned to move back the opposite way, only to see another group of about thirty civilians walking along the canal, clearly looking for them. Alpha patrol stayed where they were and when the search party got to within 100 metres, Balwanz, noticing that they were unarmed, stood up and yelled '*Kaif*', Arabic for 'get out'. He waved his rifle and they moved back but more people came up and the crowd swelled.'

Weatherford had set up the Satcom and was on the radio to headquarters describing what was happening when three shots were heard coming from the village. 'It was some kind of warning.' The Green Berets then saw a group of about twenty bedouins armed with old bolt-action rifles gathering to the west of them. Then four military two-ton trucks, a Land Rover and a bus pulled up along the highway and unloaded a company-sized unit of Iraqi soldiers. 'We have been seen,' Weatherford said into the Satcom. 'Combat is imminent. We need immediate exfil and immediate CAS [Combat Air Support]'.

'First thing we'll get you is CAS,' came the calm reply from the control officer. 'We'll work on the exfil.'

A volley of shots fired by the bedouin militia, trying to outflank Alpha, fell close and Iraqi soldiers armed with AK-47s were starting to move up the canal towards the Americans. Balwanz resisted shooting back because, 'I didn't want to fire up the Iraqis too much before we had air support.' He gave orders to ditch all non-essential equipment and blow it up to increase mobility. 'Do we really need to dump all that?' asked Harris as the men bunched up seven of their rucksacks containing food rations, extra clothing, sleeping bags, radios, cryptographic equipment and other matériel. He didn't get an answer. Weatherford just set an M700 plastic explosive charge on a one-minute fuse, packed it into the middle rucksack containing most of the communications gear and yelled 'Go, go, go!' The patrol carried off just their weapons and one rucksack between them holding water, ammunition and a spare LST-5 radio. Balwanz led them further up the ditch as Weatherford covered them, 'making sure that the stuff was burning before I took off running'.

'Some armed bedouins were moving deep to the east to surround us,' explains Balwanz. 'A platoon-sized element was manoeuvring on the right flank some 300–500 metres to the north while another platoon manoeuvred at about the same distance on the left flank. Two platoons of regular Iraqi soldiers manoeuvring along and in the canal to the west were closing in to within 100 metres of us when the rucksacks with the demolitions charge exploded.'

Firing now intensified and Balwanz ordered his men to shoot selectively at anyone armed – a lot of civilians were still gathering around to watch the show. With a single shot from his M-16 Kostrzebski blew away a farmer who was noticed trying to conceal his rifle as he scouted them from nearby. Both Green Berets with the 203 grenade launchers opened up with half a dozen 40mm grenades on either flank at the Iraqis, trying to envelop them, and fired off another barrage of grenades at the main element of enemy soldiers moving up the ditch. Killing some forty Iraqis and breaking the flanking manoeuvre, the Green Berets quickly moved up the canal some 300 metres further east, setting up firing positions at a bend intersecting another ditch where they would make their stand.

As bullets flew overhead, Weatherford again raised headquarters on the Satcom. 'Are you under fire?' asked the lieutenant colonel. '*You can roger that*!' screamed Weatherford, firing his M-16 with one hand and holding the radio with the other. They could only tell him that CAS was still twenty minutes out.

Firing single shots or two to three-round semi-automatic bursts to conserve ammunition – and avoid casualties among the village population still gathering to enjoy the spectacle – the Green Berets kept picking off groups of three to five Iraqis who continued trying to move up their flanks. Dropping soldiers at 500 metres with their M-16s, Alpha's five school-trained snipers wouldn't let a single Iraqi get past them. Frustrated in their efforts to surround the Americans, Iraqi soldiers and militia began launching frontal assaults. Shrieking battle cries and firing streams of AK-47 automatic fire, they charged across the grazing fields and up the ditch only to be repulsed by deadly-accurate M-16 and 40mm grenade fire. During a brief lull in the fighting Terry Harris lit a cigarette. Weatherford looked at him, reached over, pulled it out of his buddy's mouth and smoked it as he got back into position to start firing again. Harris shrugged and lit up another one.

It was half an hour since the shooting had started and Balwanz got very worried as he peeked over the trench beneath the hail of bullets, training his binoculars to see more Iraqi soldiers offloading from a truck which had been flagged down on the highway. Just then he heard the hollow roar of an F-16 fighter streaking overhead. The men's spirits soared, only to come crashing down when they were unable to reach the fighter crew over the LST-5 radio. 'We could hear them raising our call sign but they couldn't hear us as we repeatedly identified our position.'

The Green Berets suddenly realized that the whip antennae they needed to communicate with the aircraft on its VHF frequency had been left behind with the ditched equipment. As the jet made several passes, one of the soldiers tried aiming the round Satcom antennae at the aircraft and communication was briefly established, but they were barely able to confirm their call sign before it broke off again. The F-16 finally dropped its bombs on a communications facility some five kilometres to the south. Thinking that was the target to which he was being directed, the pilot then roared away.

One of the team's snipers, weapons specialist Robert Degroff, pulled out the PRC 90 survival radio which Weatherford had made sure got spread around the patrol when preparing the mission back in Saudi Arabia. By coming up on the emergency frequency, his voice was picked up on an AWAC, which notified the crew of another F-16 patrol flying towards the area to tune into the new channel. The two jet fighters streaked overhead, one F-16 making a right hook in the standard combat manoeuvre just as Sergeant Charles Hopkins spotted another Iraqi Army truck pulling off the highway, unloading more Iraqi soldiers, armed with mortars and heavy machine-guns. 'It really was a case of the cavalry arriving in the nick of time,' says Balwanz. 'With artillery targeted on us we couldn't hold our perimeter.'

'Go to the moon and fly towards the sun and we'll tell you when you are on top of us,' said Degroff to the F-16 pilots over the PRC. 'Trench right at the bend,' came the lead pilot's voice two seconds later, identifying Alpha's position. 'Roger, that's us,' confirmed Degroff, who immediately started directing air strikes. 'Enemy concentration behind berm 300 metres to the north.'

'Look! the bomb broke!' exclaimed a Green Beret who had never seen a cluster bomb, when the projectile dropping out of the plane's belly opened up in mid-air, releasing the bomblets, which fell around the Iraqis manoeuvring on Alpha's right flank. The ground then shook with simultaneous multiple explosions, wiping out the entire platoon.

Hopkins leaned over the trench, observing the Iraqi mortar team getting into position, setting up their base plates as an officer gave urgent firing orders. He spoke into his PRC to direct the second F-16 on to the line of trucks. Another set of clusters rained down among the vehicles, covering the stretch of highway with explosions which turned the trucks, Land Rovers and buses into flaming carcasses, ripping up people and equipment anywhere near them. Spotting another group of enemy positioning a heavy machine-gun to the left flank, he continued: 'concentration 300-400 metres south from last hit, hit it'. This time a 2000lb bomb crashed on to the field with a deafening blast.

Guiding in every air strike in relation to the one before, Alpha called strike after strike from sorties of F-16s which came roaring out of the skies during the remainder of the day. Two fighters were almost continuously on station to cover the besieged Green Berets, spraying the fields around them with high explosive. Cluster bombs came 'danger close' to within 200 metres of Alpha's position as the planes were directed down the ditches against persistent Iraqi efforts to charge up through the canal. 'It was like a continuous earthquake, the ground wobbled all around from the reverberating explosions of countless bomblets, turning the world into a bowl of jello.'

Using direction-finding equipment, the Iraqis penetrated Alpha's radio frequency, tuning in a beacon to misguide the air strikes. When the pilots reported the problem, Alpha just responded 'bomb the beacon' and they saw a 2000lb bomb explode on a van in the far distance.

Headquarters called in, giving word that exfiltration originally scheduled for 16.30 could not take place until dark. As the afternoon wore on, Balwanz decided 'to assault back down the canal to make sure that the enemy wasn't massing just around the bend, preparing a sneak attack for when the sun goes down'. He led the way, armed with his MP-5, followed by Sergeant Robert Gardner, carrying the M-16-203 combination. Crouching low along either side of the ditch as their legs sank into pools of mud, the two men moved cautiously, step by mud-drenched step. They had covered over 100 metres when Balwanz saw some soldiers, 'an enemy point element', hunkered down some twenty feet in front. The Green Berets instantly fired and charged down the canal shoulder to shoulder, firing from the hip on full automatic as they sprayed all three Iraqis with countless rounds. Trampling over the bodies, they spotted more soldiers running away. Gardner pumped out two 40mm grenades 'to encourage them'. Following the flashes, the Green Berets moved through the smoke-filled ditch, where the smell of cordite blended with that of the wet mud. They saw blood tracks marking where Iraqi dead and wounded had been dragged off, and continued further down, coming across the remains of Alpha's ditched equipment.

They confirmed that the rucksack containing all the communications and classified equipment in which the explosive charge had been placed was totally destroyed. But much of the other stuff was still in usable condition. It was getting late and starting to grow cold. Some food and extra clothing would be nice. Balwanz and Gardner slung four surviving rucksacks, one over each of their shoulders and headed back up the ditch. They stopped to pick up the AK-47s from the three dead Iraqis on the way and called in by radio for two more of the men to help carry the loads and provide security. Some sporadic fire was being received from an unseen location down the canal.

'The sun was low in the sky as I dropped the gear back in the perimeter,' recalls Balwanz. 'Hopkins immediately told me that another fifteen to twenty soldiers had just arrived on the road and were moving into the canal. Our defence was readjusted and as another sortie of F-16s arrived, Degroff called in the CAS to again interdict the forces in the canal. This was done very effectively with 2000lb bombs.

'As night fell, the detachment prepared for another attack by enemy forces. Through our NVGs we could detect some movement but no attack was mounted. As our 8 p.m. exfil time neared we decided to move back to a safer position for the helicopter RV. Climbing out of the trench, we made a retrograde movement across 300 metres of open terrain to cross another canal and set up a new perimeter behind a berm. The F-16 crew flying cover gave us word that two MH-60 Blackhawks were twelve minutes out. Degroff vectored in the helicopters by turning the PRC-90 survival radio on the beacon mode and the MH-60s landed directly on to the LZ some metres away from us. We lifted off without incident and returned to Saudi Arabia feeling weary but very happy to be alive.' Alpha had killed some 250 Iraqis that day.

Looking back on the most dangerous day of his Army life, Weatherford admits that Alpha were very lucky that the Iraqis didn't have any armoured vehicles nearby. 'If there had been tanks that would have been it. Even light armour would have forced us to surrender.' Armoured personnel carriers were what Bravo Two Zero encountered when the eight SAS men tried manoeuvring over flat, barren desert out of their compromised position. 'The rumble of the tracked vehicle came from the south. I strained my neck and saw that on the far left hand side an APC with a 7.62 machine gun had come down a small depression,' writes Sergeant Andy McNab in his autobiographical account of life behind Iraqi lines. "Fucking let's do it! Let's do it! Let's do it!" I screamed at the top of my voice.'

Two of his men fired off their .66 anti-tank rockets and a second APC with a turret-mounted gun opened fire all along the area. 'All you are is a foot patrol and these anonymous things are crushing relentlessly towards you. You know they carry infantry and all the details about them. You know the driver's in front and the gunner's up top and he's trying to look through his prism and it's difficult for him and he's sweating away up there, getting thrown about trying to take aim. But all you can see is this thing coming screaming towards you and it looks so anonymous and monster-like, magnified ten times suddenly because you realize it's aiming at you. They look so impersonal, they leave destruction in their wake. It's you against them. You're an ant and you're scared.'

According to McNab, the eight SAS men overcame their natural instincts to just lie on the ground and 'make yourself as small as possible' – which wouldn't do much good in the desert – and charged the APCs even as another vehicle pulled up loaded with Iraqis. In destroying the truck and one APC, he estimates they killed at least fifteen enemy soldiers before managing to retreat to higher ground still under fire from the second APC some 800 metres away. It was at that point that they decided to ditch their rucksacks with all the heavy equipment and start legging it to the nearest border with Syria. Their radios were useless. Having been issued with the wrong batteries, they couldn't communicate from their depth in denied territory and no air support or emergency extraction could be called in.

During Bravo Two Zero's epic ordeal, Private Bob Cosiglio was killed holding off Iraqi forces for nearly two hours with his belt-fed 5.56 Minimi machine-gun. The diminutive five foot three trooper, who liked to spend his free evenings chatting up women in discotheques, received several gunshot wounds in his head and left thigh, but McNab managed to escape, only to be captured a day later as he got within sight of the Syrian border. Corporal Stephen Lane died of hypothermia after swimming across a freezing river trying to evade Iraqi forces near Syria. The patrol had already trekked for several days through rain and sleet in mountainous areas where hypothermia also killed Sergeant Vince Phillips after he lay in a slush-covered tank berm all day, hiding from an Iraqi patrol.

'They should have made sure to take some of their warm kit with them when they ditched all their gear,' says SAS survival expert Lofty Wiseman. 'The weather turned out to be a worse enemy than the Iraqis.' Sergeant Chris was also feeling hypothermic when he had to abandon Phillips in the snow and could have kicked himself for not taking an extra pullover and one of the quilted sleeping bags. Especially as in the midst of the battle with the APCs he had rushed back to his bergen through enemy fire to retrieve a silver whisky flask which was a gift from his wife. Chris believes that if the sun hadn't come out the next day as he and another survivor, Stan, reached some lower ground, he also would have frozen to death.

The only food they had were the emergency rations in their belt kit: two packets of biscuits. A bedouin shepherd found the two men lying in a ditch and offered to lead them to some food. Chris stayed behind while Stan followed the shepherd, only to be led into an Iraqi position and captured. The SAS's fiftieth anniversary couldn't have been better celebrated. What happened to Stan was exactly the way in which David Stirling fell into captivity when he became a starving fugitive behind enemy lines in North Africa, trusting a bedouin who promised him bread. Stan would be reunited in prison with McNab and the two others he hadn't seen since the patrol split up during the mountain trek. They were all brutally tortured by Iraqi intelligence.

Chris managed to keep going on his own, walking away from the wadi where they had met the shepherd, after Stan failed to return by mid-afternoon. Wearing the sheepskin coat which the bedouin had left behind, he could at least keep warm. He continued for two more days, boxing around Iraqi villages and military positions and generally keeping to high ground to avoid the more populated areas along the Euphrates Valley. Knowing the Syrian border was west of him, he navigated almost entirely on his compass. The maps they had 'were too small scale to be of much use'. Having only managed to refill his water bottles once with some muddy river water, he was seriously dehydrating by his fourth night on the run when he heard the wail of an air-raid siren. Observing the area around him with the night scope on his rifle, he saw all the signs of a major military complex. There were several S60 AAA cannon emplacements with sentries patrolling the perimeter and large radio masts in the background.

He tried working his way around the heavily guarded installation and by what he believed to be a miracle, came across a small stream of clear liquid flowing over white rock. He immediately filled his two water bottles. First light caught him by a road between a vehicle checkpoint and an S60 position, forcing him to crawl for cover into a culvert full of rubbish beneath the road. Inspecting his unbearably painful feet, he found that he'd lost all his toenails and that blisters along the sides had connected up into long cuts weeping pus. When he drank the fresh water in his bottle, his lips and mouth suddenly burned from the acid liquid and he spat it out violently, barely able to contain a scream. Chris was actually hiding beneath a top-secret uranium-ore processing plant and the stream from which he had filled his bottles was an outlet for its toxic waste. Having gone without water or food for nearly two days, he seriously felt like he was going to die.

But an awareness that he was very close to the border helped him to reach for those extra reserves for which SAS men are tested during selection and somehow he found the spiritual strength to stagger across the road as darkness fell without being spotted by the Iraqi checkpoint. He hobbled about during the night, shocked at one point by the thunder and flash of an explosion. He looked behind him and saw tracer rounds shooting up into the sky. An air strike was in progress against the uranium plant. In an almost delirious state he came across a barbed-wire fence piled in three coils, separating Iraq from Syria. He climbed over it, cutting his arms and legs, and kept walking. By morning he was hallucinating and passed out repeatedly with static in his brain firing off white flashes before all went dark and he awoke several times lying face down on the ground with a bleeding nose. Exhaustion and thirst were taking over and death was approaching. But just as he was running out of his last iota of energy, he reached a small farm where an Arab family revived him with some water and food. He was in Syria.

One year after his ordeal, Chris was still semi-hospitalized, suffering from the after-effects of malnutrition, kidney disorders and other sicknesses caused by his agonizing escape from Iraq. It was almost two years before he could be declared fit for service again and rejoin B Squadron to be posted to Northern Ireland.

14

THE SCUD HUNT

Roaring through the darkness at 500mph in his single-seat F-15 Strike Eagle, Lieutenant Colonel Steven Turner, commanding a squadron of the US Air Force's most elite super-jets, was on his fourth night of scanning the desert plateau unrolling beneath him. A third of the Allies' air power was now concentrated on destroying Iraq's Scud missiles and Turner was on to a fixed multiple launcher which was delivering punishing attacks on Tel Aviv. But the most advanced infrared night viewing and laser targeting technology at his fingertips could not compensate for his lack of eyes on the ground. The continuously overcast skies of Arabia's worst winter in over thirty years made any effective air reconnaissance virtually impossible and all that his planes had struck until now were decoys. Being informed that SAS jeep-mounted patrols were roaming 'Scud Alley' just south of the Baghdad–Jordan highway, Turner immediately got in touch.

Approximate grid references for the target were communicated by the F-15 Strike Eagles via Satcom to the SAS D Squadron commander positioned nearest the area where the well-camouflaged Scud launcher was estimated to be. Immediately entering the coordinates into their Global Positioning Systems (GPS), about a dozen Range Rovers and motorcycle outriders started heading in the direction indicated. With thermal night vision scopes borrowed from the most advanced British and American tank technology attached to their Milan missile launchers and .50-cal machine-guns, the SAS patrols could scan the desert clearly over vast distances in the darkness. By morning several of the patrols had reported sighting SAM missile sites and AAA batteries dug in at various points. A picture emerged of six anti-aircraft missile emplacements, each protected by a triangle of S60 cannon forming a star-of-David pattern over a wide perimeter of desert. It was the standard Soviet system of protecting a major missile installation.

But this could still be part of a very realistic Iraqi ruse for the benefit of Allied reconnaissance. Some of the Scud decoys had been found to be highly elaborate, complete with false transporter-erector launchers with their own crews and armed escorts. Sighting the various SAM emplacements, the SAS observed activity around them for the rest of the day from positions several kilometres apart. Concealed behind barely noticeable land elevations along the flat landscape, the various OPs began seeing quite a lot of people, as well as vehicles coming in and out, including fuel container trucks. There were communications masts in the background, radar, and a lot of electronic signalling was picked up. There were sentry posts and armoured vehicle patrols. SAM and S60 crews got shifted every three hours, assuring maximum alertness. It was estimated that there was at least a regiment-sized unit in the area and it all seemed to add up to a major functioning installation. The fact that they couldn't get a good look at what was underneath all the camouflage netting, which was barely visible but extended over several hundred square metres in the middle of it all, definitely indicated that there was something quite important there.

The information was relayed by burst transmissions to SAS headquarters throughout the day and reported to General Chuck Horner, the USAF chief directing the air offensive from the war room three floors below the ground in the Saudi Air Ministry in Riyadh. The entrance was off an anonymous corridor passing through a large room with officers running around frantically in chocolate-chip suits, speaking on the phone and staring at computer screens, like a stock exchange office, with even lawyers around to advise on international law and other legal aspects of warfare. Horner was impressed with the SAS intelligence. From his swivel chair next to General Schwartz-kopf at the oval-shaped table surrounded by TV monitors and maps, the Air baron picked up his bulky green field telephone to call the US Tactical Air Control Center.

The ground erupted in front of the SAS OPs that night. Lieutenant Colonel Turner's sixteen F-15 Strike Eagles broke through the cloud cover. With targets magnified ten times on the pilots' electronic helmet visors connected up to the planes' Lantirn (Low Altitude Navigation and Targeting Infrared for Night), they came in with pinpoint accuracy. The SAM batteries were immediately silenced as the aircrafts' radar-guided missiles streaked into their control trailers. Heavier bombs made sudden pink flashes, shaking the ground. One big explosion sent up a massive fireball, appearing to confirm that something very big and flammable had been hit, probably a rocket-fuel depot. There were more deep explosions, flashes, flames, masses of sparks and the sounds of something very metallic being ripped apart as 2000lb bombs fell directly into the underground bunker complex. Fire and wreckage seemed to be everywhere following the fifteen-minute raid. Intermittent explosions could be heard throughout the night as more ammunition and fuel depots caught fire.

A reinforced enemy presence in the area the next morning prevented the SAS from making a closer inspection but they could confirm that what had been hit was definitely not a decoy and irreparable damage had been done. Scud attacks on Israel lessened considerably following the SAS guided air strikes during the closing days of January.

'The combination of patrols operating on the ground with the high-tech aircraft and armed helicopters flying all over Iraq was quite lethal,' says an SAS officer in the Gulf. The USAF headquarters charted the locations of SAS patrols in Iraq on the computers and map boards in their operations room as much to know where their planes could count on ground reconnaissance as to avoid confusing the roaming special forces units with Iraqi ones. SAS headquarters were asked to provide daily reports on the exact location of its patrols down to a five-kilometre operating radius. This was a difficult order to fill because it was not unusual for the SAS commander to be out of communication with many of his patrols for several days.

A squadron of A-10 jet aircraft designed for low-level ground attack and tasked exclusively to the Scud hunt landed at Al-Jouf to be based there through the rest of the war, joining the special forces propeller airfleet of C-130 Talons, Blackhawk helicopters, ten Pave Lows and six camouflage-painted British Chinooks. The A-10 pilots, lacking much of the high-tech gadgetry of the more advanced fighters, needed to coordinate their operations with the SAS on a constant basis. 'We also got good intelligence from the A-10s on Iraqi gun emplacements, radar systems and troops on the ground,' says one British Chinook pilot.

But when three SAS Range Rovers were driving away from their observation post along an Iraqi MSR, there was almost a forerunner of the tragedy which later struck regular British troops when they were mistakenly attacked by an A-10. Stitching up the ground towards them with its 30mm seven-barrelled Gatling gun, the jet fired one of its Maverick missiles, which exploded just twenty feet from one of the jeeps. As the SAS men dived for cover, one of them spread his large Union Jack over the ground. The pilot didn't come in for a second pass. When they drove back into Al-Jouf some days later, the patrol found out who the pilot was and located him at the bar, sitting on a stool quietly sipping a drink. 'You damn nearly killed us,' said the camouflage-clad patrol leader as he and two others came up to him. The USAF lieutenant was very apologetic, offering to buy everyone a round of drinks. Special forces have their privileges. Beer and whisky, strictly prohibited elsewhere in Saudi Arabia, were available in Al-Jouf – to calm some very taut nerves.

Looking around highway bridges and road culverts, SAS mobile columns searched for the especially elusive mobile Scuds on trans-porter-erector launchers. These could be brought out at night from bunkers and hiding places along the road for a quick shot at Israel and driven back quickly into concealment long before an air sortie could locate them. Even when SAS reconnaissance teams observed the flash of a missile launch, it was difficult for an air strike to come in much before fifty minutes, by which time it was usually too late. With coordination between ground and air increasing as the Scud hunt progressed, the reaction time was brought down to thirty minutes, 'which inhibited Iraqi launches, limiting them to no more than one or two a night. It became difficult for Scud crews to keep their accuracy,' according to an SAS officer in the Gulf.

D Squadron patrols engaged in a high-risk game of tag, shadowing Iraqi convoys until warplanes came on target and bombings could be guided down with precision. In one instance a troop commander directed two consecutive F-15 air strikes on a large and heavily escorted line of trucks trailing giant cylindrical objects along the Baghdad–Jordan highway. Between the intense anti-aircraft fire and thick clouds, the bombs missed their mark. Incapable of letting such a lucrative target escape, the captain manoeuvred his half column of vehicles to within a 2000-metre range of the highway and fired off a volley of Milan rockets, to be met by a rapid hail of high-explosive shells fired by vehicle-mounted S60 cannon. 'It was difficult for our patrols to get too close to the Scuds, which were always well-protected. There was at least one mechanized division of Iraqi troops protecting the Scud sites. The effect of AA fire turned groundwards on our thin-skinned open vehicles was very devastating.'

One sergeant manoeuvred his jeep through the curtain of .57 shells spitting up dirt all around him to launch three Milans at close range. The anti-tank missiles whooshed towards a giant trailer, which erupted into flames as the jeep went into a violent turn and a giant fireball lifted into the sky. 'We probably hit the rocket-fuel container truck,' reckons an SAS officer in the Gulf. But it was considered a Scud kill. 'We may not have hit more than a couple of Scuds during the entire campaign, but it was our constant harassment which forced the Iraqis to move their launchers away from the danger zone close to the Jordanian border. That was our objective.

'It was really a matter of keeping up the pressure and hitting targets of opportunity. SAS teams continuously raided or called down air strikes on radar stations, communications facilities, observation towers or troop concentrations. Little precise intelligence was acquired on the dispositions of the missile force, however, as the Iraqi prisoners we picked up were of a very low grade.'

Two Iraqi communications technicians were taken when a group of SAS jeeps burst into a radar facility, demolishing it with machine-gun and rocket fire. As one trooper prepared to lob a grenade inside a radio hut, the door flung open, practically hitting him in the face. One young Iraqi soldier and a much older man ran out with their arms up, getting on their knees and begging to surrender. They were rushed back to Saudi Arabia by Chinook. It turned out that the seventeen-year-old and fifty-something man were only enlisted army reservists and 'were quite useless'.

'As US special forces came into western Iraq, the Scud hunt became more and more dominated by armed helicopters,' says an SAS officer in the Gulf. By mid-February Delta Force, which had been given time to prepare and be well briefed by the SAS, was flying in teams north of the Baghdad–Jordan highway into 'Scud Boulevard'. Saddam Hussein's mobile missile launchers being chased out of Scud Boulevard were hiding along the deeper wadis between the barren ridges rising off the gravel plateau towards the Euphrates Valley, still within striking range of Israel.

Pave Low pilots had mapped their way around the SAM sites and learned to avoid the Roland batteries, whose quick-acquisition radar scared away SAS Chinooks. The helicopter pilots found that as the ground grew ever more uneven its contours made it easier to get their bearings. An armed MH-60 Blackhawk which now regularly escorted support missions followed the Pave Low's flight path, reaching a

Delta Force position shadowing a Scud launcher near the wadi of Al Qaim during the first week of February.

The helicopter crews had been on the ground for about twenty minutes, unloading fresh supplies, when they and the Delta leaders were suddenly transfixed by the flash of a Scud lifting into the sky some twelve miles away and causing an earth-shaking blast as it did so. Four Delta men immediately jumped through the lifting ramp of the Blackhawk as its blades whipped up and the helicopter took off.

Skimming the ground at its maximum speed of 140 knots, the Blackhawk soon reached the wadi, surprising the Scud crew with streams of machine-gun and rocket fire as a second missile was being mounted. Iraqi AA crews manning nearby vehicles struggled to level their S60 cannons downwards as the helicopter slowed and dived towards them. Spotting the gun emplacements through their NVGs, the Delta men hosed them with their side-window .50-cal machine-guns and their M-16s. The pilot aimed directly at the Scud outlined on his infrared sights, shooting off a dozen high-explosive shells into the rocket booster. The desert glowed as the Scud disintegrated in a red-yellow flash and the launcher was enveloped in a ball of flames.

It was one of the few fully confirmed mobile Scud kills during the special forces' war in western Iraq. The missile which was fired from near Al Qaim had hit Israel's northern port of Haifa that night. A group of Blackhawks guided in by Delta is reported to have taken out sixteen Scud launchers being prepared for a devastating blitz on Israel a few nights later. 'But they all turned out to be decoys,' according to a high-ranking special forces commander in the Gulf. What is undeniable is that the missile attacks on Israel did cease after mid-February.

With total air dominance achieved over western Iraq by mid-February, as enemy anti-aircraft radar became seriously degraded, American and British special forces officers on JSOTF's small planning cell were seriously considering establishing an advance helicopter base or FARP (Fuel Armament and Resupply Post) in Scud country. Response time to any further missile launchers would be greatly reduced if a full helo squadron could be permanently close at hand. C-130 Talons were already conducting low-altitude mid-air refuelling operations some eighty miles into western Iraq, regularly servicing helicopters coming and going from targets well north of the Baghdad–Jordan highway. At their normal operating altitudes of

250–500 feet, where they were vulnerable to small-arms fire, the Talons would just fly right through the search radars or 'blue-grey systems' picked up on their electronic warfare scanners, managing to avoid the odd 'red' ones, usually from small mobile AA batteries still cruising along the roads which could be detected at six-mile distances. 'But it was very scary at our operating altitudes – even AK-47 fire could prove devastating,' says one of the pilots.

'The area along Syria was still too heavily defended,' according to an American SOS lieutenant colonel flying Pave Lows into Iraq from Turkey. During a mission he saw one of the last Scuds fired at Israel overflying his helicopter. 'There was too much enemy activity around there to take further action. I just reported the approximate launch position to an F-15 CAP and turned back.'

An entire Iraqi brigade started hunting A squadron's Group 1 in a long chase through the desert which decimated a full half column of vehicles. Heavily armed APCs manoeuvred to cut off their southern escape routes to Saudi Arabia. Corporal David Denbury, who during a raid on an Iraqi communications base, had driven his motorcycle into an enemy position, killing several men at close range, now sped north with his half group, breaking through their encirclement towards the Baghdad–Jordan highway. Another four Range Rovers drove west under close pursuit by the Iraqis. One jeep with a troop sergeant major described by SAS historian Tony Geraghty as 'a type known to armies since the Roman Legions, a rocklike presence that was its own guarantee that everything would be OK', had been damaged by a land-mine and after fighting off two Iraqi attacks was virtually destroyed. The sergeant himself was badly wounded in his left thigh, smashed through by a 7.62mm bullet. Bleeding badly, he had to be carried by the two other men as they were forced to continue their escape on foot.

A tourniquet applied by the medical specialist to the sergeant's leg failed to stem the blood flow for long. They soon ran out of morphine to kill his pain and as he lost his strength he asked to be left behind. Tears welled up in the eyes of one trooper, who took out his 9mm pistol and asked the sergeant if he preferred to be finished off with a bullet in the head. 'Don't be stupid,' he replied. 'Save the bullet to defend yourselves. I'll bloody die here soon anyway.' The two men zipped him up in a sleeping bag, resting his head on a rucksack, and walked away heartbroken as the sight of him disappeared behind a

haze of sand in the windswept plain. They were eventually spotted by
a helicopter which delivered them to Al-Jouf, where one of the men
presented his worn-out sheepskin coat as a special gift to 'The
Mayor', Lieutenant Colonel J.V.O. Weaver.

Another six SAS men, one of them severely wounded in the
stomach, also had to abandon their two vehicles after being hit by
cannon fire. They hijacked an Iraqi civilian car and drove back to
Saudi Arabia. Three men pursued in another jeep used their PRC 90
to signal an AWAC, which vectored in a Chinook to extract them just
a few miles from the Jordanian border as Iraqi vehicles closed in on
their damaged jeep stuck in the middle of a minefield.

The remainder of A Squadron continued to operate along the
Baghdad–Jordan highway, knocking down microwave relay towers
which boosted the signals of Saddam Hussein's underground land-
lines. They often carried out the task within view of civilian traffic in
broad daylight. 'If the Iraqis had flown helicopters we would have
been terribly limited,' says an SAS officer in the Gulf. 'But since there
was no air threat, the main emphasis was to get away from
approaching ground troops as quickly as possible rather than trying
to hide.' Motorcycle outriders served as mounted sentries, scouting
the perimeters around where groups operated or were laagered up
with jeeps, circled around in wagon-train fashion. 'We didn't even
bother to put up camouflage netting over the vehicles.'

During the terrifying chases through the desert's table-like gravel
plateau, punctuated by the flashes of heavy automatic gunfire and
artillery bombardments as Iraqi armoured vehicles repeatedly came
close, Denbury had a premonition of death. Gathered up with his
mates at a lay-up point in a shallow wadi one night over a small
cooking fire hidden between jeeps, he handed over his silk escape
map, wrist-watch and other personal effects, asking that they be given
to his family and girlfriend back in Wales.

Iraqi pursuit units were soon composed of highly trained commandos who tried springing ambushes on the SAS. Enemy snipers
were already in position one night as Denbury, with a *shemag*
wrapped over his face and neck to protect him from the wind and
driving cold, peered through NVGs, scouting out the area around
where the jeeps had formed a protective circle. There was the sudden
crack of a rifle shot and he flew off his motorcycle with a single bullet
lodged in his chest. He screamed out that he had been hit as more
gunfire started breaking out, and another outrider, barely five foot

seven tall, managed to carry the six-foot corporal back to the main position. Having been pierced through the heart, Denbury died in minutes. The SAS managed to again break out of the attempt to encircle the enemy and the dead body was picked up the next day by a Chinook helicopter which was signalled down on its way back from a resupply mission further north. Another man wounded during the night's engagement, which had left a Range Rover badly damaged, was also flown out.

Yet another form of casualties were discussed at the 16 February meeting of the NCOs' mess, eighty-seven kilometres inside Iraq, at Wadi Tubal. After coming to a decision among themselves, the sergeants advised their headquarters and Squadron commanders to officially relieve three captains in western Iraq from their troop commands. 'Like in any other unit, there are people who don't perform up to standard when the bullets start flying,' says one former SAS commander, explaining a quick decision to comply with the request. The officers in question were apparently no longer exerting effective control over their troops. 'He was found not to be fully comfortable with his weapons,' was the explanation given in one of the cases. An unconfirmed version is that the officer objected to the kneecapping of Iraqi prisoners of war in an effort to extract information from them during the frantic search for Scuds.

* * *

The NCO who was left to die in the desert, to be much mourned at Wadi Tubal, where they listed him as dead, actually pulled off a resurrection. He survived long enough to be found by an Iraqi patrol, who got him to a hospital where they performed surgery on his leg. He was released together with other POWs shortly after the war to join future mess meetings in Hereford. The Regimental Sergeant Major who called the quorum of thirty NCOs in Wadi Tubal had his own problems with weapons. The Falklands War veteran lost his M-16 rifle, which jolted out of his hands as his Range Rover was nearly overturned by an exploding shell during 'some drama' with Iraqi tanks. Delta's only fatality, Sergeant Pat Hurley, died in a helicopter crash as he was being evacuated from Scud country with a leg wound.

15

THE DECEPTION

A squadron of SEALs and some Kuwaiti sailors were enjoying a barbecue of freshly caught fish on a giant barge sitting off the coast of Kuwait when they looked up at the passing lights of the mighty Allied air armada flying above them. The opening days of the war came as something of an anti-climax for the SEALs, who had been among the first units to reach the Gulf, digging into the Saudi border with Kuwait within the first week of Saddam Hussein's invasion.

Recently relieved of their desert duties after 'staring at Iraqis through binoculars for six months', Lieutenant Riley's platoon had refitted the big creaking barge, working together with the Kuwaiti sailors, who had driven it out on their escape. They had mounted machine-guns, rocket emplacements and a helipad on the floating dock which they named 'Sawhill Three', turning it into a platform for special operations in their 'natural sea environment'.

The SEALs were not the only ones supplementing their foul-tasting military rations by fishing. They had been observing a team of Iraqi special forces holding seven offshore oil drilling platforms not far away, dropping plastic explosive charges into the water to net their daily catch. The day after the air offensive, the SEALs made final preparations for their opening shot, which had been planned for a long time. As evening fell, Lieutenant Riley moved Sawhill Three to within 1000 yards of the oil rigs, casting off with his men on Zodiac boats motoring up in pairs with muffled outboard engines to observation points beneath the drilling platforms. He opened radio contact with a half squadron of eight Army Apaches hovering out of the skies and with another SEAL platoon choppering off the frigate USS *Nicholas*, which cruised just over the horizon.

Approaching their targets, the Apache pilots, looking through their thermal-imaging night-vision cameras, reported sighting a total of twenty-three Iraqis moving on the decks of five rigs. Scanning the area below with their NVG-compatible night scopes, the SEALs spotted no enemy activity in the surrounding waters or understructures. Flying low passes over the rigs, the Apaches shot up the platforms with 30mm cannon fire. They had orders not to use their Hellfire missiles as it was imperative to minimize damage to Kuwait's oil industry.

The Iraqis quickly surrendered. By the time the SEALs clambered on to the installations from their Zodiacs and roped down from helicopters, the twenty-five surviving commandos stationed on the separate oil rigs in groups of five or six, had all come out on to the decks with their arms up. Five on board one platform who tried shooting back had been killed by the strafing Apaches. 'They all had parachute and diving patches on their uniforms indicating that they were special forces,' recalls Riley, who found the oil rigs well stocked with night-vision equipment, gas masks, shoulder-fired SA-7 missiles and light AK-47s with collapsible stocks. The Iraqi commandos had been providing an early-warning screen against approaching Allied warships and aircraft.

For the men of SEAL Team One, trained along the resort beaches of Coronado Island off San Diego, California, it was all 'kind of matter of fact'. What for many was their first combat experience seemed less hard than most practice missions. Commander Holden, who had operated in the Gulf with the naval task force escorting Kuwaiti tankers through the Straits of Hormuz during the Iran–Iraq war, spent a lot of time observing the Iraqis through television pictures relayed back to his command trailer by an RPU. This model-type aircraft mounted with an infrared video camera could be flown by remote control over enemy positions in Kuwait providing 'real time displays of activities behind the lines'.

The naval commando war geared up when the SEALs were tasked with de-mining the Gulf waters after a US missile cruiser was badly damaged by an Iraqi acoustic mine planted further out than previous mine-locating operations had indicated. Guided by the sonar signals from fast-moving frigates, SEAL scuba teams were dropped into the water from helicopters, diving into the lower depths of the Gulf to blow up twenty-six mines with underwater demolitions charges. Laying aside their oxygen tanks and picking up MP-5s, a platoon flew on to capture the elusive Iraqi mine-layer which had been immobilized by missiles fired from an A-6 Intruder. Again the crew put up no resistance when the SEALs roped on board.

SEAL reconnaissance operations along Kuwait's coastline became more frequent and got closer to the shore as D-Day approached. 'It would be real ugly if we had to take the beaches,' says Riley. As they probed up to within 150 yards from the Kuwaiti coast in their Zodiacs, the SEALs could see the highly fortified beaches with five lines of barbed wire extending into the sea in a basket-weave pattern. Land and water mines were planted between the holes with interlocking fields of fire from bunkers protected behind wire-mesh fences to detonate anti-tank rockets before they could hit home. 'Saddam's military engineers were crafty.'

Ensign Tommy Dietz, known as 'Iceman', had got used to seeing enemy mines floating round his Zodiac when on his fourth night he silently slipped over the side and into the cold blackness of the water, swimming to the shore with two other SEALs. To hide the rubber shine of their wetsuits, they wore green pilot's overalls zipped on top. Their faces were painted black and dark green under their diving masks. Emergency radios were wired under their garments, providing communication with each other, with the men left back in the Zodiacs or their rigid 'cigarette boats' farther out at sea. While most SEALs took MP-5s, a swim team would include one or two support shooters armed with CAR-15s, miniaturized collapsible versions of the M-16, with longer-range stopping power, fitted with 203 grenade launchers.

'We swam until we could stand on the bottom and keep our heads just above the surface. Taking cover behind obstacles, we just observed. It was quiet. The Iraqis had no patrol boats out. We spotted sentries moving around on our infrareds but were totally confident that they wouldn't see us. Everything we saw we recorded by whispering into small tape recorders. After half an hour we swam back.'

In fact, there was never going to be a big amphibious landing on Kuwait. But it was in the interests of the Allied high command to keep the Iraqis believing that Marines would come crashing on to their shores as in *The Sands of Iwo Jima*. It would keep them off balance and their forces dispersed. In one of its most ingenious manipulations of Gulf War media coverage, the American military's Joint Information Bureau arranged a marvellous press extravaganza, flying out hundreds of journalists to cover a division-sized Marine beach-assaulting exercise on a Gulf island off Abu Dhabi. The event was very deliberately promoted as a realistic practice for the upcoming battle of Kuwait and at a time in which everybody was waiting for the war to start, it made dramatic film footage to fill in vacant air time on CNN. The show was also closely watched by Iraqi generals in their command bunkers.

As far back as the previous October, senior officers of the US Marine Expeditionary Force had proposed staging a feint landing on the Kuwaiti coast supported by naval gunfire while their main force moved across land from Saudi Arabia. The trouble was that the threat of mines would make it dangerously risky to bring large ships close to the shore and the Marines discussed the plan with the SEALs. On 18 February Ensign Dietz was called into Commander Holden's headquarters and told to prepare 'the concept' for the operation. Dietz formulated a plan in consultation with his squad, presenting it within hours. The 'tasking' papers came back four days later. Dietz had barely time to line up the explosives, other equipment and stage a brief rehearsal with his men before the 'execution' order came within the next twenty-four hours. General Schwartzkopf had brought back the date for the land offensive. The twenty-seven-year-old SEAL, who had been in the unit for three years, had an hour to finalize his squad's preparations for their special role in the liberation of Kuwait.

Holden walked on to the pier of Ras Al Mishab that night as Iceman Dietz and his eight-man squad boarded their cigarette boats. The commander wished them luck and with rare military formality the eager young ensign snapped him a brisk salute before jumping on to the wide-framed thirty-three-metre racing boat, casting off its ropes and skimming out over foaming waves, raising clouds of spray against the blue moonlit sky. Three other 'cigarettes' followed.

Also called 'fountain boats', these extremely fast and highly manoeuvrable special warfare vessels 'are a cross between an aircraft and a submarine', say the SEALs, because when their three-man crews drive at their normal speed of 80mph 'you are either flying or under the water'. The SEALs requisitioned them from a fleet of drug-running vessels impounded by the US Coast Guard in Florida. They were originally custom-made by a British boat builder catering exclusively for Caribbean narcotics smugglers with specially designed hulls which can resist speeds of up to 120mph. The SEALs modified the boats by coating the hulls with bullet-resistant Kevlar and adding radar arches, GPS systems and gun mounts for heavy machine-guns, rockets and light cannon.

'The only trouble with 'em,' a SEAL mechanic points out, 'is that the Navy doesn't stock spare parts for their high-performance racing engines, so they have to be openly purchased on the civilian market and that's expensive.' The SEALs could hardly refer to their sinister, racy, new delivery vessels as 'drug boats' but since maritime smuggling traditionally involved tobacco, they adopted the term 'cigarette'.

Although the orders had come in at the last minute, Dietz couldn't have planned a better night for a covert beach landing. 'The sea wasn't too rough and we had a perfect half moon, giving us just the right combination of light and darkness. Ideally you want a moon to backlight the beach when you are swimming into it.' After approximately an hour's journey sixty miles up the coast, the cigarette boats reached the grid reference point recorded on their GPS, some ten miles off the beach of Mina Al Saud, just south of Kuwait City. The eight SEALs unstrapped their two inflatable Zodiacs from the bows, laid them in the water, deposited all their equipment on the wooden board platforms between the rubber canvas sides, piled on board and cast off into the black, choppy seas. The long silhouettes of the four mother ships which would remain on station slowly disappeared behind them.

When their Zodiacs were within 500 metres of the beach, Dietz and five other SEALs slipped into the water, whose temperature was in the low fifties. Each towed a small haversack containing 20lb of C-4 plastic explosive in two sticks fitted with a timer and a detonation cord. Being on a combat mission, each man also carried a small day pack strapped to his waist and containing a chocolate ration, water and extra ammunition – enough to keep them going for twenty-four hours. Side-stroking gently with one arm while holding up their silenced weapons with the other and kicking beneath the water with the help of fins fitted over zip-up rubber boots, the SEALs glided in with the current.

Getting close into shore, they dived beneath three coils of concertina wire with shallow water contact mines floating around them. Fortified bunkers loomed into view on the beach ahead as the SEALs crawled up to the water's edge, washing up with the gentle waves. The six men spread out twenty metres apart, lying still for a few minutes just partly submerged in six inches of water, taking cover behind the various sea obstacles as they unclicked the safety-catches of their sub-machine-guns. Satisfied that there were no enemy patrolling nearby, they spread further apart, to about fifty metres, and went to work attaching their haversacks with the explosives to whatever barbed-wire entanglement or steel obstacles were closest. They set the timers on the charges to two hours and hid them under the wet sand. Slowly crawling to within easy distance of each other, they eased themselves backwards into the surf.

By the time the tired SEALs piled back on to the Zodiacs, laying down their weapons, pulling off the masks from their faces with running camouflage paint and zipping out of their dripping-wet flight suits, it was just past 01.30. All along Kuwait's southern border, the Allies' massed divisions of tanks and mechanized infantry were beginning to move as last-minute checks were finalized for the long-expected land offensive scheduled to roll in at 03.00. The outboard engines of the two Zodiacs were pushed back in the water and with a muffled roar they turned and headed back to the cigarette boats.

Two speedboats were still waiting to take the squad back to base, while the other pair, loaded with another SEAL squad, headed slowly up to shore. At 02.30, they turned on full speed and rushed up to the outer lines of barbed wire on the beaches, savagely firing broadsides of .50-cal machine-gun fire and rocket grenades on to the shocked Iraqi defenders. Within the next quarter of an hour, the beach erupted as six separate 20lb charges of plastic exploded along a quarter-mile stretch at two-minute intervals. Word swept down the Iraqi front that an amphibious invasion was starting and two divisions stayed in place along the coastline with their guns pointed out to sea. Dietz was already drinking hot coffee back in his tent at Ras Al Mishab when he got news at 0.600 that the big deception had worked. 'I'm glad we saved lives.'

It was time for US Marines of the 1st Force Reconnaissance Company, who train closely with the SEALs at neighbouring bases in southern California, to come out of their holes. Since the start of February, five-man teams of the special reconnaissance unit known as 'the guys who eat rocks' had been operating on the ground several kilometres inside Kuwait, observing Iraqi movements, to open a path for the 1st Marine Division, which would come dashing up the coastal highway to Kuwait City.

When the Marine armoured columns started rolling over the border in light M-60 tanks, the 1st Force Recon Company were racing ahead in their new Light Strike Vehicles, known as Fast Attack dune buggies. Like Delta Force and the SEALs Marine recce units had recently been issued with the new Chenowth fast attack vehicles originally built for overland racing, winning the 1990 Baja California 1000 competition.

Enabled by their vehicle's race-track suspension to take poundings three times their weight, the Marines, wearing motor-racing helmets, flew their LSVs over the sand dunes at top speeds of 80mph, maintaining course and balance as they landed in clouds of sand over hidden Iraqi positions.

The LSVs were the first vehicles to come tearing into Kuwait City's corniche. Racing down the sea-front boulevard between looted luxury high-rises and the rear of Iraqi beach-side defences, they shot up soldiers fleeing from the bunkers, hosing down machine-gun and grenade fire as ecstatic Kuwaitis came on to the streets to cheer them on. Meanwhile, Marine Corps M-60 tanks were briefly detained just south of the city, finishing off an Iraqi armoured brigade caught between the airport and the beach of Mina Al Saud, which had remained so heavily guarded after the SEALs' visit the previous night.

The British SBS helicoptered on to the roof of the British Embassy along the corniche to raise the Union Jack and clear the building of suspected booby-traps. In a more secret operation an SBS team was inserted by submarine along south-eastern Iraq's marshy coastline to observe the explosive situation developing around the port of Basra. The local Shia population was rising up against Saddam Hussein as the Republican Guard became cornered there by the main thrust of Schwartzkopf's armoured columns moving in a flanking manoeuvre from the west.

At the other extremity of the front, CRAP platoon of France's 2 REP had been landed by helicopter to set up OPs at several points along the Euphrates, within 100 kilometres of Baghdad. They were dropped ahead of the French Rapid Deployment Daguet Division driving up from Saudi Arabia to take the airbase of As Salman, linking up with elements of the American 101st Airborne Division. As Salman was found deserted by the time French and American paratroopers started pouring through its bombed-out aircraft bunkers with their thick concrete shells perforated by gaping holes from laser-guided 2000lb bombs. CRAP also missed getting into good fire-fights. But intelligence they could supply to Schwartzkopf was very heartening – there were no sizeable Iraqi troop deployments between the Allied positions and Baghdad.

16

BODYGUARD

At his high-rise apartment in Alexandria, Virginia, overlooking the placid Potomac River, Commander Richard Marcinko switched off his television as General Norman Schwartzkopf declared the liberation of Kuwait and ceasefire negotiations with Iraq. The US Navy SEAL counter-terrorism specialist did some last-minute packing, took the only beer left in his refrigerator, drained the can then crushed it, checked his watch and sat down to wait for the Federal Marshal who would be coming in a few minutes to take him back to jail. The time limit on his temporary furlough from a three-year prison term had expired. A second call had never come from the intelligence officials who had approached him behind bars about doing some deep-cover work in Iraq, so it wouldn't be extended.

Marcinko had originally been sentenced on charges of misappropriating government funds, for some minor personal expenditures, and taking kickbacks from a private consulting company set up by a former subordinate in SEAL Team Six. Many in the SEAL community believe that he was framed by a clique of senior naval officers who didn't want to see him being promoted to captain and eventually to admiral. The system, it seemed, was being a lot harder on Marcinko than on Saddam Hussein.

In the early days of the Gulf crisis, Richard Nixon, whom George Bush had once served as an ambassador, had publicly spoken about 'taking out a contract on Saddam Hussein'. There are reports that Israel's intelligence service, Mossad, was seriously working on the project and that some elements in US intelligence departments felt it imperative to get participation if not control over any such operation. Was a super ex-SEAL undercover assassin being considered to spearhead a final solution to Saddam? There is nothing to indicate whether such an operation never got off the drawing boards in either

Washington or Tel Aviv because the war moved too fast, or because it was deemed impracticable or as a result of certain high-level compromises between the powers concerned. An effort to personally target Saddam's family and inner circle was ultimately left up to a woman member of Kuwait's fledgling resistance movement.

Safaa was among the very elite of Kuwait. An American-educated woman investment banker and niece of a former ambassador to Washington who loves Las Vegas but could not bring herself to abandon her native land when the Iraqis invaded. 'I felt like I was leaving my heart and my soul behind,' she says about the moment when she had packed all of her bags and was ready to catch the flight to London which her father had arranged. 'I just couldn't breathe and stayed where I was.' Appearing to carry on with her life as usual from the luxury of her walled villa, Safaa befriended influential Iraqi officers. She also got to know Muhammad Dashti, a robust, quick-witted Persian with deadly green eyes who had the habit of curling the sharp points of his jet-black moustache as he ran around with a 9mm pistol tucked in his trousers. His job was directing the Abu Basil resistance group operating under the authority of Sheikh Salem Al Sabah, a cousin of the Emir and Kuwait's Interior Minister in exile.

There were few who knew about Safaa's secret activities except Dashti and an Iraqi major who collaborated with them out of personal hatred for Saddam Hussein's Baathist regime. The major assigned to security duties in Kuwait City repeatedly warned Safaa whenever they were trying to intercept signals from the satellite telephone she kept hidden in her house in order to communicate important messages to others outside Kuwait. One night when Iraqi soldiers did come to search her villa and had Safaa cornered in her bedroom in only a night-gown, the major raced over to ensure that she came to no harm.

As was her custom before the war, Safaa continued making business trips to Baghdad. 'It was very easy now that Kuwait was part of Iraq for Kuwaitis to go to Baghdad. We didn't need passports.' She secretly took videos of resistance activities and Iraqi atrocities in Kuwait to an undercover intelligence operative at a foreign embassy who smuggled them out by diplomatic pouch to the BBC in London. 'It was of critical importance for the world to know that there was an active opposition against the Iraqis inside of Kuwait. That our people were willing to give their lives for their freedom.' One important film she delivered depicting an Iraqi tank being blown up by a bomb thrown on top of it by a group of masked men (who included a well-known Kuwaiti footballer) as it passed under a highway bridge 'was worth more than destroying a hundred tanks'.

Safaa would stay at Baghdad's Babel Hotel, where, on 27 October 1990, she met Uday Hussein, Saddam's notorious and obese son, who was making a lot of money selling Ferraris and other luxury cars which he drove out of Kuwait together with large quantities of gold. Uday befriended the shapely, long-haired Kuwaiti and through him she got to know officers of the Iraqi Mukhabarat as well as several high-ranking government officials including Cabinet-level ministers who regularly congregated at the Babel's bar for afternoon drinks. Sitting in their company Safaa immediately got the idea of planting a bomb, which she could smuggle in inside her handbag, to wipe out some of her country's leading tormentors.

She proposed the operation to resistance leaders when she got back to Kuwait City, and demolitions experts, including some unidentified foreigners, immediately started building a compact device loaded with the amount of plastic required to do the job. But when Safaa returned to Baghdad for a final recce, she found that highly trained presidential security men had strengthened vigilance in the Babel, checking everyone coming in and out. With the assistance of some other sympathetic members of the Mukhabarat, at least one other effort was made to smuggle an explosive device into a high-level ministerial office. But as a result of Saddam's air-tight system of interlocking cells of bodyguards in which one group keeps an eye on another, the plot was discovered and at least one man was shot on the spot.

During the time that Saddam Hussein was considered a virtual ally of the West, during his ten-year war of attrition with Iran, American secure communications systems, British infrared night vision equipment and French radar-guided missiles, were among the many items he was allowed to acquire to enhance the capabilities of his mainly Soviet-equipped army. In the late 1970s Saddam also imported the best Western know-how to bolster his personal protection services.

Between 1978 and 1980 a five-man team of recently retired sergeants from the SAS were contracted to go to Baghdad to train the Iraqi presidential bodyguard to 'bring them up to the standard required of official protection services in western countries'. The

Iraqi Mukhabarat had been employing trainers from Germany's GSG9 but was dissatisfied with them, cancelling their contract after someone was killed in a helicopter rappelling exercise. The 80–100 Iraqi 'untouchables' who composed the dictator's personal escort were all carefully recruited through Saddam's family and close relations in his northern home town of Takrit, and were sometimes required to perform a murder to prove their loyalty. The most privileged and feared men in Iraq, often with authority over Cabinet ministers, Saddam's protectors were extremely arrogant and very difficult to train, convinced as they were about their place among the elite of the elite – the inner core of their national leader's praetorian guard.

'We had to start out by proving our ascendancy over them because they thought that they did not need training and certainly not from Westerners,' says one of the British trainers. This was done by constantly humiliating them, slapping the men around and soundly beating them in individual karate matches in front of others to humble Saddam's chosen few and force them to accept instruction. They needed to be schooled in the basics of crowd surveillance and selective shooting, especially in public places, to effectively protect their man when he came out among his people – something which Saddam liked doing to retain his charismatic hold over a significant proportion of Iraqis.

Before their SAS training, Saddam's bodyguards would have been likely to open up with sub-machine-guns on an entire crowd if they sensed an assassin attempt. By using 'killing house' techniques with mock crowds made of pneumatic dummies, the British team conditioned them to discriminate between hostile targets and non-threats, how to direct instant accurate fire along a very narrow circumference upon one or more armed individuals popping up among many who are unarmed. They were also taught French-developed methods of 'protection de rapproche', protecting a VIP from a distance, standing away from him in a crowd instead of closely around him – something about which Iraq's, if not the world's, most challenged bodyguards had no idea. The ex-SAS men also advised the planning of Saddam's security. Observing his daily routine, they redesigned Baghdad's streets, suggesting the removal of central dividing walls and even pavements along avenues which Saddam regularly travelled so that his armoured Mercedes could always make a quick U-turn in the event of an ambush.

Once the five ex-SAS men were satisfied that they had prepared and trained one of the most dedicated and professional official protection services in the Middle East, with few rivals anywhere, they delivered their final report to the respective authorities and quickly left Baghdad. There were no fond farewells and they never planned to return. 'Such were the trials and the humiliations we put Saddam's bodyguards through to get them up to the level required for their job that we were honestly afraid they would turn on us. To reach our standard in shooting and unarmed combat we had to make them hate us and they could easily have some revenge planned. That's the kind of blokes they were. The training mission in Iraq was strictly a one-off.'

It was Saddam's bodyguards who were undoubtedly the ones inspecting Safaa Jafar's handbag when the Kuwaiti Mata Hari was thinking of planting a bomb at the Babel Hotel. Saddam was turned into such a hard target that he amazed the world soon after the Gulf War with television pictures of him coming out to greet the crowds on the streets of his bombed-out capital. At a time in which half his country was rising against him, his army was vanquished, the most elite American, British and French forces were poised in their helicopters a short distance away and the atmosphere was rife with intrigue, coup plots and roaming assassins, the world's public enemy number one was shaking hands and kissing babies on the streets of the Iraqi capital, while his crack security men stood by anonymously in plain clothes. It was an impressive if ironic testimony to the professionalism of SAS personal protection training.

17

'GET NORIEGA'

Ever since Aaron Bank's abortive plan to get Hitler, special operations undertaken by the USA, Britain and France tend to stop short of one strategic objective in warfare: the elimination or capture of government and military leaders of the opposing side. The SAS cancelled their plan to kidnap General Menéndez in the Falklands. However extensively special forces penetrated Iraq, no concerted effort was mounted in the end to eliminate Saddam Hussein. Sergeant Mick of the SAS remembers his squadron being stood up to go into Libya in 1969 to unseat Colonel Gaddafi. 'We are going to give that fellow Gaddafi a good duffing,' his squadron commander told them. 'It was marvellous,' says Mick, recalling how they all assembled to go storming on to an airfield in Tripoli. But they never got past the aircraft hangar of a British base and decades of Libyan-supported terrorism were endured instead.

The international legalities and complexities of the diplomatic deal-making which usually accompanies so many current-day conflicts might even require special forces to occasionally protect an enemy. Perhaps history has come full cycle from total war to limited war under the pragmatic rules of the 'new world order'.

But it was the very political circumstances which President George Bush had to consider as he contemplated US military action in Panama during 1989 which made the Latin American republic's rogue president, General Manuel Antonio Noriega, the primary target of Operation Just Cause. Conceived in its computerized code-name as Operation Gabel Adder, the plan was originally strictly limited to a fast, low-key special forces action to kidnap Noriega and fly him back to face drug-trafficking charges in the United States. Delta, SEAL Team Six, sixteen special forces helicopters and an electronic intelligence team, backed up by a company of

Rangers and two AC-130 gunships, were going to do the job as originally proposed by General Luck, the then commander of Joint Special Operations Command.

Panama was a country with which the USA was not at war and had been a virtual US protectorate since the turn of the century, with large American military bases controlling the trans-isthmian canal. There was an active and organized opposition movement to Noriega although this had been recently repressed as the self-proclaimed 'maximum leader' started cancelling elections. Most importantly, perhaps, the civilian population was well disposed and friendly to the USA, and it was imperative to avoid the unnecessary casualties or 'collateral damage' which would inevitably result from an all-out invasion. It was a mission tailor-made for special forces.

President Bush himself had once had friendly dealings with Noriega when he was CIA's director and the Panamanian colonel was the country's useful intelligence chief, drawing a retainer from Bush's department comparable with the salary of the President of the United States. Noriega's drug-trafficking activities had been known for some time. Bo Gritz maintains that when he repeatedly called attention to them during his time as commander of the Panama-based 7th Special Forces Group (7SFG), he had been unequivocally told, 'lay off this guy'. But as the 'war on drugs' became a priority for the Bush administration at a time when Noriega's ties with Colombia's Medellín cartel were growing, the criminal enterprises of a once-valued intelligence asset could no longer be overlooked.

While tons of cocaine passed through Panama and billions of dollars were laundered through its banks, filling Noriega's accounts, 7SFG teams based in Panama were being flown out to fight the drug wars in Bolivia, Colombia and other South American countries where SAS training teams were also employed. Green Berets patrolled rural cocaine-producing areas, reporting back to Drug Enforcement Administration (DEA) officials at the respective American embassies about the activity of local drug barons. American and British special operatives trained local security forces on how to mount surveillance of cocaine laboratories and drug airfields and destroy them.

US pressures on Noriega to leave the Panamanian presidency, even the offer of the carrot of safe passage to a retreat in Spain, got nowhere, perhaps because of the Panamanian's fears that if he relinquished power in any deal with the Americans, the Medellín cartel would kill him. 'Pineapple Face', as his old CIA handlers called

Noriega because of the deep acne scars disfiguring his large head, increasingly resorted to anti-American rhetoric to justify his moves towards dictatorship, even turning to Cuba for military support and protection. Operatives from Castro's secret police, the Dirección General de Inteligencia (DGI), were brought in to train and arm Noriega's most loyal units in the Panamanian Defence Forces (PDF). They also advised on setting up the 'Dignity Battalions', gangs of armed thugs used to break up opposition rallies and harass democratic leaders. Noriega even fell back on Castro at one point to help him settle a quarrel with Colombian drug barons.

Operation Gabel Adder was held back as efforts persisted to secure Noriega's voluntary resignation. But the Panamanian dictator only used the time to entrench his personal control over the PDF and promote his most loyal officers into key positions. When a coup attempt against Noriega did take place in October 1989, the USA failed to back it because of uncertainties about the reliability of officers planning the putsch. It soon became clear that the situation in Panama was too far gone for Noriega's personal removal to change matters substantially since he would in all probability just be replaced by other colonels whose integrity and commitment to democracy was dubious at best.

There was also the question of American intelligence and military officers who themselves might have been compromised by Noriega. Not only were handsome pay-offs making their way into certain American-controlled Panamanian offshore accounts, but, according to the US Ambassador to Panama at the time, Everett Briggs, Noriega regularly resorted to sexual honey traps. 'He typically invited distinguished American visitors to his residences or hideaways and placed them in exquisitely compromising positions. This was done through the presence of willing and aggressively available beaùties, often beauty contest winners of all ages and descriptions, who presented themselves in an uninhibited environment.' Once they became participants in Noriega's orgies, 'the officials would be video-taped and photographed'. One US Army colonel attached to the planning group of the Joint Chiefs of Staff says that he was once invited to a party at a Noriega residence during a trip to Panama and entered the room to find Noriega sitting naked on a couch surrounded by four nakedly voluptuous teenage girls. He was invited to join them but walked out instead.

'The critical point of this issue,' states an internal State Department memo written by a prominent US ambassador in Central America, 'is not the admittedly salacious one, but rather the implication that Noriega may have compromised enough high level US officials to thwart US policy and buy US inaction toward his regime in Panama.'

It was evident to the Pentagon and the Bush administration that to effectively remove Noriega's regime would require the dismantling of his entire army, intelligence and internal political apparatus. Such an expanded goal could only be accomplished with a much larger military operation than originally envisaged for Gabel Adder. But special forces would still be performing the central role in the plan 'built around the three principles of maximum surprise, minimum collateral damage and minimum force'. Serious preparations began with the replacement of the long-time commander of US forces in Panama, General Woerner, with the more aggressive General Thurman, who immediately requested the deployment of additional forces. A contingent of 2000 extra 'military police' arriving at US bases in Panama to deal with the heightened tensions in October 1989 actually included units of Green Berets and one squadron from Delta Force. Two squadrons of Pave Low and Blackhawk helicopters were also flown in secretly with their blades and tails folded up on board giant C-5 Galaxy transports.

'Pre-invasion training started on 5 October, two days after the failed coup and continued until 18 December, practically into H hour,' says Sergeant Duane Stone of the 7SFG, who was among those recently flown down from the USA. 'We were based at a World War I coastal gun battery living in a cement-covered bunker as we underwent intense training of twelve to eighteen hours per day seven days per week. Some on-ground recon of the targets we were given was also being conducted. One of them included the Modelo Prison fortress, where Noriega kept political prisoners.

'If we had to move 200 yards to put a gun into action carrying 175lb loads it was done and gone over realistically until we got it right. In one case we couldn't fast-rope down from a helicopter carrying that quantity of equipment, so the ammo was packaged in a honeycomb container and dropped from twenty-five feet. Snipers practised moving ahead of the main force with lighter loads for pre-determined distances between the helo drop and target.'

'After the failed coup we intensified training to refine our capability to get Noriega out of power,' says Colonel Robert Jacobelly, commander of US Army Special Forces in Panama, 'although we still did not believe that anything would come of it. We didn't think that President Bush would have the will to invade Panama over the scattered harassment incidents and drug activity.'

'We had to put up with a lot of shit from the Panamanians,' according to another Green Beret, Sergeant William Merces, from Texas, who joined from the Rangers and had been based in Panama for several years. 'There was constant harassment from police and officials always stopping us for no reason while we were driving. They always asked for bribes. Just getting from one side of the city to another would cost about thirty dollars.' The Americans were under strict orders not to start any fights. 'When serious training really began I was really looking forward to doing what we had been trained to do.'

Delta was primarily charged with keeping a surveillance on Noriega with the aid of a specialized team of electronic listeners. Delta and SEAL Team Six would be mainly responsible for his capture and rounding up his main supporters, accomplices and cronies throughout Panama City as well as the country's second largest town, Colón. There was another key operation assigned to Delta which grew in importance as relations between Panama and Washington deteriorated. A CIA communications specialist, Kurt Muse, working under cover setting up a clandestine radio network for the democratic opposition, had been arrested and jailed by the PDF. He was being held in solitary confinement in a prison across the road from Noriega's headquarters, the Comandancia. His rescue was imperative as Noriega was threatening to kill him. Muse was especially vulnerable as the Comandancia would be the first target hit in the invasion plan.

Delta started preparing a rescue with information obtained from a Red Cross doctor visiting the imprisoned CIA agent. On the basis of his observations, Delta was able to compose a computer simulation of the approaches to Muse's cell. White walls outside the building gave way into an interior corridor flanked by brown walls on the coloured computer screen, with distances calculated by squares criss-crossing the floor. The image on the screen would move up four flights of stairs, with walls and sides going white again. Past a landing the picture then moves down a corridor, turning left, then right, then through a door. Muse's cell door is on the wall opposite in brown and the guard is sitting behind a white desk with a set of big, heavy keys hanging from the wall. Using the computer blueprint, Delta was able to devise an insertion, penetration, room-clearing and extraction plan which was rehearsed repeatedly on a full-scale plywood model built at a location in Florida.

Delta also put the Comandancia under permanent surveillance, as well as Noriega's main residences, including a ranch house in the interior of the country around Río Hato. SEAL Team Six would take down Noriega's beach house and his alternative headquarters at the PDF base of Fort Amador near the Canal. Helicopters, C-130 Talons and Specter gunships continuously rehearsed their various phases of the operation, including approach routes and landings on their various targets in Panama City and the area of Río Hato. While the Blackhawks and Pave Lows familiarized themselves with their flight paths by flying over Panama City, the Specters rehearsed their main opening assault on the plywood replica of the Comandancia built near their home base of Hulburt Airfield. It was estimated that Operation Just Cause would require some 200 aircraft flying at low altitudes during the night and day over the tightly constricted airspace over Panama City.

Securing Noriega's capture also required closing all of his possible escape routes. The main airfields, including Tocumen-Torrijos airport in Panama City and the airbase of Río Hato, would be immediately occupied by special forces and Rangers in the first few minutes of H hour. The SEALs were tasked with neutralizing the beach-side airfield of Patilla, where Noriega kept a private Lear jet and a Puma helicopter. It had to be observed and approached from muddy coastal flats where the SEALs would land by boat and make their way around the runway into buildings where they would set up a .50-cal sniper rifle to disable Noriega's plane from a distance of 900 metres. They were under strict orders to minimize any damage to the aircraft. SEALs swimming underwater would disable PDF patrol boats. Air strikes and artillery had to be kept to a minimum and an underwater approach, attaching magnetic limpet mines to the gunboats' hulls, was preferred in order not to risk damaging Panama's fishing fleet, which was docked close to them.

The increasing tensions between American servicemen and the PDF came to a climax in mid-December with the shooting of a US Marine and the arrest of a naval officer and his wife, who was sexually molested. It was a provocation which President Bush could not tolerate and he immediately ordered Just Cause to be set in motion. 'I had just picked up my son and daughter on the evening of the seventeenth who were visiting for Christmas when I got a call from headquarters saying that I had to go to work,' recalls Colonel Jacobelly. 'We were ready to go on Monday night of the eighteenth because they knew exactly where Noriega was,' recalls one special forces officer. But last-minute preparations weren't yet completed and H hour was postponed until the following night, by which time Delta had lost Noriega.

Major Higgins of the 7th Special Forces Group had recently returned from one of his training missions in El Salvador and was on the runway of Howard Air Force base outside Panama City at 3 a.m. on Monday night, loading his bags on to a C-130 destined for Bolivia. He checked his watch, wondering why Jacobelly hadn't arrived yet with the final brief on his drug interdiction mission. Higgins walked back to the main building, lifted a phone and called headquarters. 'Don't you know what's happening?' came Jacobelly's voice from the other end of the line. The biggest drug interdiction operation of all times was about to take place in Panama. Higgins went back to the plane, took out his bags, loaded them into his car, and drove for the next hour and a half back to his 7SFG base.

'It was abuzz with activity with a lot of determined-looking faces as guys pulled their gear out of wall lockers. Weapons and night-vision equipment got drawn out from the arms room. Maps were being set up and equipment was getting loaded,' into MH-60 Blackhawk helicopters which discreetly ferried the 7SFG's third battalion to their staging point at Albrook Airfield throughout the day. They assembled there inside an aircraft hangar, undergoing final preparations and receiving last-minute briefings. At 19.00 hours, soon after it was dark, four-man teams started moving out with their heavy backpacks, radios and loaded weapons to take up hidden OPs around various targets in Panama City: Cuartel Cimarrón, which served as the barracks of Noriega's elite Batallón Dos Mil, and Cuartel Tinajitas, with its heavy mortars, as well as two key bridges which would isolate Panama City and the main broadcasting station. 'We heard the firing start at 00.25 and finally realized it was for real,' recalls Higgins, who went outside to have a look as some of his men took up defensive positions in drainage ditches around the hangar. 'Through the darkness, we could see the flashes in the distance, around the Comandancia.'

Sitting in the darkened sensor suite of his Specter Ghost, looking at the hazy green picture on the low-light TV screen, Captain Schneider saw the windows along the sides of buildings some 7000 feet below him blow out as his 105mm howitzer rounds pierced the roof and exploded on the third floor. Noriega's headquarters complex was subdivided into thirteen buildings. The four Specters, circling above the Comandancia like vultures at altitudes between 6000 and 10,000 feet, were each targeted on three buildings programmed into their fire-control computers. Two rounds were shot into each building in the first few seconds. 'We were like a bunch of surgeons cutting up a body perfectly,' says Schneider.

Inside a van pulling up near the Comandancia ahead of Army mechanized units driving in from the Canal Zone to surround the compound, Sergeant Dan Jones, the US Air Force Special Forces Combat Control officer, helped to direct the Specter's fire against hidden anti-aircraft positions. 'Specter 01, this is Gator. Fire Mission. Target number eight. Troops in fortified AA position, do you have my beacon? Over.'

'Gator, Specter 01, have your position and target standing by. Over.'

'Roger, Specter 01, fire with 105.'

'You will see four rounds exploding. Confirm whether target is neutralized.'

'Roger, target is burning at this time. End of fire mission.'

Of two dozen Chinese-made double-barrelled AA 30mm cannon protecting the Comandancia only two ever got to fire back. All of them were destroyed in minutes.

There was only one building in the Comandancia complex being spared the Specter's fire: the small prison across the street where two minutes before the shooting started a light MH-6 helicopter had landed on the inner courtyard. Six men leaped out carrying MP-5s and one shotgun, then ran through the open door, up four flights of stairs then through several corridors, turning right and left a few times before bursting into the guard room in front of Muse's cell. As the sentry looked up, a silent triple burst of 9mm rounds pierced his forehead. Three men covered the doorway to the corridor as the captain and two others moved up to the cell, took the set of keys hanging from the hook on the wall, opened the steel door and grabbed the stunned-looking Muse. Quickly making their way back to the staircase as they heard the sounds of Panamanian soldiers coming up from the floors below, they ran up the remaining flights of steps to the roof, where the MH-6 was hovering down to land. The rescue operation had taken a little over the three and a half minutes achieved during final rehearsals. The Specter's 105mm howitzer shells were beginning to crash down as Delta Force jumped on to the small helicopter with their rescued hostage.

The MH-6 was taking off when a spray of automatic rifle rounds fired from a range of less than twenty feet by Panamanian soldiers running out on to the roof, pierced through its windscreen and thin armour. Sparks flew as the bullets damaged the controls and the aircraft lost power, punching a hole in the building as it slid down along the wall and came crashing down into the street. Miraculously, no one inside sustained anything worse than cuts and bruises. The team sergeant emerged from the overturned wreck, leading the way out as others unstrapped themselves from their seats. One of the rotor blades which was still turning whipped around and hit him in the head. Saved by his Kevlar helmet, he ducked the second blade and laid down covering fire on to the roof as the five other members of his team, Muse, the pilot and co-pilot huddled around him. The group then ran down the street past a PDF sentry who just gaped at them open-mouthed. Rushing for two blocks, they came across the point elements of the mechanized battalion pulling in around the Comandancia. 'Bulldog!' shouted the Delta officer, giving the arranged password as soldiers escorted them down another block, over a wall and into safety. The Delta officer who led the way in the Muse rescue was awarded the Silver Star.

'It was like looking into a Roman candle,' is how Major Skip Davenport describes his blacked-out landing on the Río Hato airfield north of Panama City. The concentrated tracer rounds from PDF AA batteries, explosions, flares, burning vehicles and buildings had a blossoming effect on his NVGs, dilating his pupils as he brought down a C-130 Talon loaded to its maximum capacity with 150,000lb of jeeps, motorcycles, fuel, ammunition, rockets and as many Rangers as could squeeze in. He was manoeuvring flash-blinded through the battle zone at a speed of seventy-five feet per second.

There were parachutes blowing around everywhere which had been left lying around by Rangers who had initially jumped in to secure the airfield. One canopy caught up in scrub encroaching on the small, badly kept runway and got sucked into one of the C-130 engines, forcing Davenport to shut down the whole right wing of his Talon as it skidded around, its tail scraping against a tree. There was a lot of heavy fire coming in from the nearby PDF barracks. Suddenly an explosion violently shook the aircraft as 'an artillery shell hit right off the right wing tip. Everybody thought that we had taken a direct hit.'

Positioned by the runway watching it all was Sergeant Wayne Norrad, who had jumped into Río Hato with the 3rd Ranger Battalion, setting up a perimeter defence around the airfield. 'We jumped from 500 feet and I landed very quickly because of my large equipment load,' which included his Combat Air Controller's three heavy radios. 'There was a lot of confusion in getting to the airfield. A radio transmitter broke, we had to go through 200 yards of tall grass and I had to lead the way for the Rangers because I was the only one with NVGs.' There were explosions everywhere as a Specter gunship tried to hit two PDF APCs moving on to the landing zone 'but I thought it was coming from the Panamanians'.

The Rangers landing with a Delta team at Río Hato were assaulting the headquarters of Noriega's 'Machos del Monte' special forces battalion and his ranch house, which formed part of the complex. The Panamanians, however, were prepared and putting up a stiff resistance. The Rangers had sustained fifteen casualties. The Machos had been alerted by a Stealth bomber which was supposed to have destroyed the barracks just before the assault, in a conventional air strike which some USAF generals had insisted on. But the supersonic fighter designed to hit missile silos deep in the Soviet Union ended up dropping its 2000lb explosive load some 500 metres off the target, waking everybody up. There was a bright flash and a crashing explosion as Norrad finally got on to the runway and turned to see a PDF APC turned into flames from a Specter 105mm shell. He also saw the incinerated car and bodies of a PDF officer and his wife and children, who had pulled into the airfield as the assault began, being hit by a Ranger anti-tank rocket.

From the runway, Norrad began directing Specter fire on PDF positions as Rangers closed in on the barracks with the help of the machine-gun-bearing jeeps unloading from Skip Davenport's C-130. They wore infrared arm bands so the fire control officers in the Specter could distinguish friend from foe. Norrad began 'engage building Hotel' using the target's code. 'We've got snipers in building four can't get to them take it out . . .' To make room for five more C-130 Talons landing at Río Hato that night, Davenport managed to pull his damaged plane off the runway in a three-engine take-off.

Meanwhile three SEAL platoons crawling through deep stretches of mud and silt between their Zodiac landing point and the Patilla runway, had been ambushed by the PDF and suffered serious casualties. The operation to neutralize Noriega's private aircraft had gone drastically wrong. At least twenty highly trained and well-concealed PDF snipers were protecting the airfield which provided Noriega's main escape route. Radio intercepts picked up by electronic listening teams working with Delta indicated that Noriega was heading towards Patilla during the initial minutes of the invasion. He hadn't been found either at the Río Hato ranch or at his beach house raided by SEAL Team Six. Instead, he was at a PDF brothel very near Tocumen Torrijos airport when the invasion began. Rushing out and into an unmarked car followed in convoy by dozens of bodyguards, Noriega had managed to clear the area, heading for his private jet before the Rangers landing at Torrijos could set up roadblocks.

The SEALs' orders had changed at the last moment. Their original plan to disable Noriega's plane with .50-cal explosive bullets fired at the engine or fuel tanks was abandoned when on their approach to Patilla they were suddenly told to get into the hangar and slash the tyres. Someone, apparently, didn't want any damage done to the Lear jet.

H hour had been brought back by fifteen minutes from the originally scheduled time of 00.45 and gunfire was going up all over Panama City by the time the SEALs started approaching the hangar. According to a special forces officer listening to the action in Patilla from the secure room at the Pentagon, the changed orders meant that instead of slowly creeping on to the airfield behind the treeline along its perimeter, the SEALs now had to make a dash across the exposed runway. 'It was more of a conventional military assault for which the SEAL force was not large enough,' says an officer acquainted with the plan. The result was that snipers bunkered down inside the hangar behind cement-filled fuel drums and a machine-gun position on one of the roofs decimated the advance squad of SEALs becoming exposed on the runway within thirty metres of the hangar. One SEAL was instantly hit by several rounds of automatic fire, a second fell dead immediately after him, while a third was badly wounded. The officer leading the assault crawled helplessly between the bodies isolated from the rest of his platoon left further behind, who could only hug the tarmac trying to cover themselves as best they could against the deadly accurate PDF fire spraying up the ground towards them.

SEALs believe that the Panamanian force were equipped with night-vision equipment because as soon as the assaulting platoon's chief petty officer exposed himself slightly while aiming to fire a grenade from his M-16's 203, he was instantly riddled by high-calibre bullets. An entire squad of eight men were all dead or wounded in five minutes and as the second platoon tried to reinforce them and cover their withdrawal another SEAL was hit. As if enough wasn't going wrong, air support from a Specter gunship circling above Patilla could not be called in because radio contact had been lost. The third platoon which remained hunkered down along the airfield's flanks could only communicate via their direct Satcom link to the Pentagon. Refusing offers of emergency extraction, they only called for a medevac to pick up the five wounded, one of whom bled to death while waiting. Using their .50 rifles, the SEALs gradually picked off the PDF defenders, keeping the airfield covered throughout the night. Noriega finally didn't make it to his escape jet, being turned away from Patilla's main gate according to some reports because of the continuing fight, and his special guard force at the airfield abandoned their positions before the morning, leaving three dead behind.

Colonel Glynn Hale, a Ranger officer on the planning staff for Just Cause working out support arrangements between special forces and elements of the 82nd Airborne Division flying in to reinforce Panama City, remembers that the SEALs were 'very vague about their exact plan in Patilla' when he discussed it with them. 'They were winging it,' he says. 'I was worried about them.' According to Hale the SEALs would not agree on any prearranged ground link-up or fire support plans, convinced that they could carry out the mission on their own. They preferred to leave any support arrangements on an 'on call basis only'.

Major Higgins was watching the flashes going up around the Comandancia in the far distance when he was tapped on the shoulder by his radio man running out of the hangar at Albrook. The OP stationed on the hill overlooking Cuartel Cimarrùn had just reported an eight-truck convoy heading out towards Pacora Bridge and into the centre of Panama City. If Noriega's crack Batallón Dos Mil, which had intervened decisively to break the coup in October, was able to reinforce the city, PDF resistance would be a lot tougher to reduce, involving prolonged house-to-house fighting in drawn-out urban guerrilla combat, the kind of scenario which the planners of Just Cause had laboured to avoid. Higgins glanced at his watch. It was just after 00.30. The Rangers were only starting to land and had their hands full at Río Hato and Torrijos airport. Batallón Dos Mil were obviously well organized, having mobilized and reacted a lot sooner than expected. Higgins made a quick decision. He would have to block Pacora Bridge with the only twenty-four Green Berets who could be spared at that moment.

Immediately ordering his two A teams to load up with full combat equipment into each of two Blackhawks, the slender, fair-haired Green Beret commander jumped on to the lead helo as it took off and made a beeline straight towards the bridge, flying directly over Panama City. Lines of tracer were coming up everywhere. 'I've got something to tell you,' said the pilot to Major Higgins as the helicopter overflew the shallow river. 'The convoy is already on the LZ.' Higgins strapped on his NVGs and looked out through the windscreen, seeing the eight trucks in the green haze coming up to the far end.

The blocking force would now have to head off the bridge crossing from only the near end and Higgins gave orders to land on the other side. The two Blackhawks made a hook over the road, flying on top of the convoy and hovered in to land on the opposite bank. Tall elephant grass undulated under the hurricane blast of the turning rotors kicking up dirt and mud everywhere as the men poured out carrying 70lb loads with claymore mines, grenades and M-16s fitted with 203s. They instinctively ran towards the road about 150 metres away, taking up positions some fifty metres from the bridge as the first PDF truck started rolling on to the far end. Sergeant McDonald immediately fired his three LAW rockets, sending the truck crashing off the road and on to the bank. The other men pumped out a hail of grenades from their 203s and fired off a bipodal drum-fed 5.56mm machine-gun, its rounds ricocheting off the bridge's steel railings and making distinct sparks in the darkness.

During the forty-five-minute fire-fight Staff Sergeant Don Bowman suggested to Major Higgins getting as close as possible to the bridge and placing claymores across it. Receiving his CO's approval, Bowman moved up to the road closely covered by a sniper and placed three mines facing the entrance to the bridge. Panamanian machine-gun fire stitched up the ground all around him, displacing one of the mines as Bowman took cover at the side of the road. With thick covering fire laid on for him, Bowman crawled back on to the road and readjusted the claymore. Just then, he noticed shadows coming over the bridge's curved middle which became three men charging at a low crouch, wearing gas masks and firing AK-47 rifles. Bowman rolled across the road, squeezing off an incessant automatic burst from his M-16 as he got out of their line of fire, hitting one of the Panamanians, who fell dead just a few feet from him. The covering sniper shot another one, who fell over the side. More kept charging up from behind.

'Get the fuck out of there,' Bowman heard someone shout. 'Specter is going to fire.' In the next instant an AC-130 gunship coming overhead decimated the charging column of the Batallùn Dos Mil with magnesium rounds from its 25mm chain gun, covering the bridge with thousands of razor-sharp steel flechettes with skin-piercing fins, making a whirring sound as they ricocheted everywhere in a galaxy of sparks. Bowman could see one shadow crawling out through the cloud of smoke, struggling, still managing to move despite his destroyed limbs, then finally stopping dead.

The Green Berets started moving across. One Panamanian suddenly tore through them on a motorcycle, firing an automatic pistol. His wheels were shot out and he was taken prisoner. Grenades were thrown over the side of the bridge in case there were snipers and demolitions teams hiding underneath. A defensive perimeter was taken up on the far side, and scattered shots were exchanged with remnants of the Batallón Dos Mil throughout the night. The persistent sound of the running engine of a PDF truck was picked up on the heat-seeking sensor of the Specter circling overhead. The outlines of the truck's body appeared as a dark smudge on the infrared screen with a clear white light on its front end. It was there that the gun control officer placed a dot quad activating the Bofors gun, which automatically fired a 40mm shell straight through the hood of the truck, destroying the engine while leaving the rest of the vehicle intact.

Large quantities of ammunition, heavy machine-guns, mortars, anti-tank weapons and recoilless rifles were recovered from Batallón Dos Mil's wrecked convoy at Pacora Bridge when the Green Berets and 82nd Airborne Division paratroopers fanned out to search the area the next day. It would have been enough material to drag out a guerrilla struggle for the control of Panama City over several weeks. All nineteen captured Panamanian soldiers were wearing civilian clothes.

After hiking a mile loaded down with 100lb of equipment through fifteen-foot jungle grass, and crossing a stream, Sergeant Ramón Cantu, one of the many Hispanics in the 95 per cent Spanish-speaking 7SFG, finally got to his OP position at the top of a 300-foot hill. Fighting was already breaking out throughout Panama City. Cantu focused his binoculars on the Cuartel Tinajitas across the valley just in time to see its 120mm mortars rolling out in open trucks. Before he could report their position – the moisture was eating up his radio batteries – the three vans had dispersed into nearby neighbourhoods. Only an occasional flash could be spotted as the mortars hurled shells into the American bases along the Canal Zone and the Ranger positions at Torrijos airport throughout the night and into the next day.

Noriega's pre-recorded voice calling on Panamanians to resist the Yankee invasion bombarded the airwaves from the PDF broadcasting centre in Cerro Azul. A Green Beret team fast-roping on to its roof put it out of action with the assistance of a telecommunications technician who dismantled the antennae. The Army wanted to save the station to use it in the next few days to broadcast appeals in support of the US choice to replace Noriega, President Endara. They wanted the station totally undamaged. The few PDF sentries offered no resistance but Noriega's voice kept coming from one or two other radio transmitters hidden around the city.

Meanwhile Noriega himself was nowhere to be found as he kept moving from safe house to safe house, through the homes of supporters, friends and trusted acquaintances, managing to stay one step ahead of the special forces teams hunting him down. SEAL Team Six burst into his offices at Fort Amador to find eight million dollars of drug money piled up in notes inside his safe. Captain Woolard, commander of Team Six, roped down with forty men 'from a Pave Low helicopter on to the deck of a ship lying off Colón harbour which, it was believed, could be used for his escape, but found only the crew and some of his known cronies who were taken prisoner.

'We were working through a blacklist of 80 to 100 individuals who Noriega was expected to contact and turning up unexpectedly at their addresses,' explains a special forces officer. Permanent surveillances were set up around the Cuban Embassy and other diplomatic missions where Noriega might take refuge. At one point a special forces team burst into the Nicaraguan Embassy when it was thought that he had gone in there. Specter gunships and Pave Low helicopters were continuously kept on station flying over his possible sanctuaries. Many places on the target list were whorehouses, residential homes and apartments belonging to mistresses, political collaborators, businessmen, lawyers and PDF officers.

Dressed in jeans and trainers, with nylon windcheaters or denim jackets covering their pistol-packed shoulder holsters, a SEAL Team Six squad ran up the steps to the second-floor apartment of Dr Núñez. After blasting open the front door with a sawn-off shotgun, they poured through the flat, the flashlight on an MP-5 suddenly shining on Núñez lying in bed with his girlfriend. She sat up, covering her breasts with the sheets as they yanked him out at gunpoint, clamped on handcuffs and carried him off. One of Noriega's body-guards would turn up there some hours later to be told that the Americans had already been around. He would either rush back to tell his leader that the place was unsafe or possibly decide to abandon Noriega and save himself. During his first few days as a fugitive Noriega managed to remain hiding at some locations for as long as ten hours awaiting the reports from bodyguards he had sent to scout out other locations. As the hours and days went by, fewer and fewer of them returned.

'Hey, slow down, bitch!' exclaimed the C-130 Talon tanker pilot over his radio at an inexperienced female crew member flying a giant KC-135 supersonic tanker some 30,000 metres above Panama. 'You're going too fucking fast, get off your jet speed so I can juice up. I've got all kinds of guys down there calling for support.' There was a million gallons of gas in the sky for aircraft coming into Panama. After refuelling helicopters throughout the night, the special forces Talon was down to its minimum of 3000lb of gas and he had some customers waiting below. The KC-135 got the message and the Talon was finally able to come up from below, insert its probe into the fuel hose, replenish with some more tons of gas and head back down to refresh some thirsty Blackhawks and Pave Lows.

Some one hundred helicopters, ten Specter gunships and other aircraft were buzzing over Panama City as morning broke on 20 December. 'Panama was close quarter combat,' says Captain Tom Trask, comparing its tightly constricted urban airspace with the limitless desert over which he later operated during the Gulf War. 'It was much scarier in Panama.' There weren't instant acquisition anti-aircraft missile radars to worry about but 'you could actually see the guys shooting at you coming up from behind windows and trees aiming their AK-47s'. One Blackhawk pilot, Captain Porterfield, located a sniper firing at his helicopter from a crane and shot him with his window-mounted machine-gun. Special forces helicopters pumped out grenades from 203 launchers into windows and rooftops, 'burning out' sniper and machine-gun nests.

After initial night-time operations around Torrijos airfield and Modelo Prison, where many potential hostages were liberated, Trask's Pave Low and Porterfield's Blackhawk were put on the Noriega hunt, following radio reports coming in from special forces teams sniffing him out on the streets. They carried Ranger reaction teams and followed all incoming reports. '. . . Girl says he's headed to apartment complex between Balboa and Oceano Street . . . turn right then left at intersection after Holiday Inn . . . Roger, we're heading there.' The helicopters would chopper over with snipers taking positions at the side window while others got ready to rope down as vehicles screeched up below. 'But we were just chasing ghosts.' A lot of the contacts being rounded up by Delta and SEALs were deliberately feeding false leads as decoys were also used to mislead special forces. A car radio Noriega had been using during the initial hours of the invasion could no longer be picked up. 'Noriega lacked a secure communications network,' says one special operative, 'preventing him from directing an organized PDF resistance. But he had an escape and evasion plan already worked out through his sophisticated network of contacts operating around Panama City.'

As a Cuban freighter started moving through the locks of the Panama Canal closely watched by Marines and regular Army troops, a Specter gunship was ordered to keep circling over it, focusing its low-light TV camera on the ship's deck in case Noriega tried to jump on board. One mobile PDF AA battery managed to keep firing well into the morning of the twentieth, shooting sixty-round bursts at special forces aircraft as they flew right over it before hiding among the alleys and thick clumps of trees which covered much of the Panamanian capital. Despite damaging one Specter on its way to support 82nd Airborne Division troopers moving to secure Americans trapped in the Marriott Hotel, the battery was never found.

'Our sensors picked up snipers on a rooftop position firing on our troops approaching the hotel,' says Captain McCruthen, whose Specter had taken some hits from the elusive AA gun. 'We hosed them down with our 40mm and were then ordered to move on to take out snipers firing on American military dependants at the residential quarters of one of the bases. After we put down some rounds there, they quieted and we flew on to locate a heavy 120 mortar shelling Army headquarters at Fort Clayton. It took us a while but we finally found it and took it out.'

'One of the biggest fears,' explains Colonel Jacobelly, 'was that PDF units would mount a concerted mortar or rocket attack on Howard Air Force base, where we had loads of aircraft and fuel blevits lined up right next to each other along the runway. It would have caused absolute havoc. A lot of the Rangers and Green Berets were used to clear the perimeters around Howard and other sensitive installations. They took out several PDF platoon-sized elements caught in the process of setting up mortar and RPG positions. Some were neutralized just on time.'

Jacobelly adds, 'We also failed to realize the degree of civil disorder which ensued in Panama City. It sucked up a lot of forces.' 82nd Airborne Division troopers landing to set up key roadblocks around the city and occupy strategic buildings and facilities to support the Noriega hunt and other special operations soon had their hands full dealing with the massive looting and widespread chaos engulfing the capital as fighting progressed throughout 20 December and order broke down completely. 'We expected most Panamanians would just stay at home and wait for the fighting to be over. But it didn't happen that way.' There were incidents in which panicky American paratroopers inadvertently shot Panamanian civilians. Tanks from Army mechanized units shelled residential neighbourhoods in the process of clearing out pockets of PDF resistance around the destroyed Comandancia.

The more specially trained Green Berets and Rangers being relieved by conventional forces were needed in the interior of Panama to take PDF garrisons and prevent a resistance movement from developing in the country's remote jungles and hills. Barring a few sniper incidents and token attempts at resistance by local PDF commanders, the pacification process in rural areas was achieved bloodlessly. Green Beret A teams supported by Ranger platoons would usually land by helicopter at a town or village holding a PDF barracks or *cuartel* and call up the commander from a local telephone, asking him to surrender. If he refused, they called in a Specter gunship to give a demonstration with its 40mm Bofors gun. 'Each and every time there was a pretty fast decision to turn over control of the location to us,' says Major Higgins, who went on to conduct the interior operations directly after his stand at Pacora Bridge.

Green Beret Sergeant Corbin remembers 'the toughest part of my Panama assignment' as trying to convince the lieutenant commanding a Ranger platoon 'to change his attitude. That not everybody speaking Spanish was an enemy' and to stop coming on so heavy towards the civilians as they patrolled through some southern villages. Members of Noriega's Dignity Battalions were active there, telling local Indian villagers that 'white men painted in green would be coming around to kill them. So I told the Rangers to wipe the camouflage paint off their faces, take off their helmets and look friendly . . . something very difficult for Rangers to do.' At one point Corbin and the lieutenant almost came to blows but the Green Beret finally convinced him that if he didn't do it his way 'we will just be leaving a trail of booby-traps behind us'.

The Ranger lieutenant realized that a more relaxed approach was working when, trying out Corbin's advice, they immediately started to pick up better intelligence and cooperation from the local people. They learned about narcotics shipments coming in from Colombia through the coastal town of La Palma near the Río Hato area, being shown into the walled villa of a local drug baron and close associate of Noriega, where they found suitcases full of cocaine and an arsenal of twenty AK-47s. Inside the luxurious living room, equipped with expensive stereo, video and colour TV, was a large picture of him posing with Noriega. There were lots of wild parties, the Americans were told, with a lot of women, loud music and guns fired into the air late at night. The local Noriega 'straw boss' had fled the area shortly after the nearby Machos del Monte headquarters had been taken.

As day three of the invasion progressed, systematic raids, round-ups and arrests conducted by 150 men of Delta Force and plain-clothes SEAL Team Six commandos rolling up Noriega's network in Panama City continued relentlessly through the day and night. But Noriega still couldn't be found. Ninety Team Six operatives house-hopping and searching boats through the port of Colón weren't having any better luck. Helicopters hovered at rooftop level over gutted buildings while American tanks, APCs and jeeps patrolled the rubble-strewn streets encountering mobs of rioting Panamanians looting and snipers shooting. The occasional RPG would also come hurling towards US vehicles.

But PDF heavy mortar activity had practically ceased and more peaceful crowds of Panamanians were starting to turn out welcoming the American soldiers, cheering them on for overthrowing the hated dictator and offering food and drink. Only some neighbourhoods still heavily filled with armed Dignity Battalions remained off limits for all except armoured vehicles and heavily armed special forces teams. Noriega's bellicose radio speeches calling on Panamanians to repel the Yankee invaders persisted on AM and FM frequencies, encouraging the hopeless resistance efforts of the most fanatical among his followers.

It was at about noon that Sergeant William Merces, who remained at the hangar in Albrook with a small 7SFG reaction force, was informed that a transmitter broadcasting Noriega's speeches had been traced to a privately owned broadcasting station operating out of the top floors of a skyscraper in the centre of town. He was ordered to immediately destroy the facility. While a special forces team made their way by vehicle, Merces flew over on a Blackhawk to link up with them by roping down on to the roof. After sliding down forty feet over Panama City's skyline he reached the antennae as the men entering from the ground raced up through the building's eight floors. The large, stocky Texan started placing demolitions charges around the thick base of the steel tower but realized that he hadn't brought enough explosives to knock down the antennae. When some of the others met him on the roof, they could only supply him with a few hundred grams of plastic packaged in frame charges. Sabotage was not an operation which special forces teams were prepared for in Panama, where the emphasis was always on minimizing damage.

Merces just headed for the station's interior. Going down the fire-escape to the seventh floor, he breached the emergency door, proceeded down the corridor and kicked open the door to the recording studio. Nobody was there. The only sound was that of Noriega's pre-recorded rantings on automatic replay. Merces levelled his M-16, shooting up the radio equipment but Noriega's speech continued. He then placed a square-shaped M700 demolition charge on the FM system, fused it with a ten-second timer, stepped back out of the room and watched it explode, spreading a fire which consumed the AM transmitter in the next few seconds. There was finally silence. 'I really enjoyed that,' says Merces.

* * *

The eight-man Delta team which had just leaped out of two blue Fords were bursting into their fourth whorehouse in as many days. Shrieks went up from the ladies-in-waiting as the armed men stormed in through doors and windows. The soldiers drew down their pistols and MP-5s, not wishing to shock the half-dozen girls lounging around in bikinis and black lace underwear. The big-breasted head madam froze up as the chief sergeant came up to her holding his .45. He produced a crisp fifty-dollar bill, slipped it down her cleavage and she began to talk. 'He was here just a quarter of an hour ago.' She was too shaken up to lie. The sergeant was led up some stairs with three of his men, through a dimly lit corridor and into Noriega's private suite. They instantly recognized the distinctive odour of his Havana cigars, seeing a lot of fresh butts piled up in the ashtray on the night table. The bed was rumpled, and passing their hands over it, the Delta men could feel it was still warm. He had somehow given them the slip again. But at least they could now be certain that he hadn't managed to get over the nearest border to Costa Rica as some of their senior officers were starting to think.

In the tropical New Year's afternoon Captain Don Timpson was flying his Specter above Panama City. The fighting had died down but fires, looting and general civil disorder persisted throughout the isthmian capital. Clouds of smoke rose into the sky, casting a brown haze as the rising sun began to shine brightly. Urgent orders crackled over his radio headset, telling him to fly on to the residence of the Papal Nuncio. Timpson checked his computerized charts showing all of the different targets and waypoints around the city. Nothing had been entered for the Papal Nuncio. 'Where's that?' 'It's across from the Casino and next to the Holiday Inn.' Now he knew where it was; he had overflown it several times. 'Roger, I'm on my way.' He turned the controls to manoeuvre his AC-130 around in a semicircle and flew on for a few minutes to stay circling above the large three-storey house surrounded by tropical gardens where Noriega, with one last loyal bodyguard, had just turned up requesting sanctuary. He had finally run out of places to go.

Paratroopers were soon jumping off jeeps and APCs as they set up roadblocks and surrounded the Papal Embassy. Delta teams also stood by discreetly in unmarked cars and vans as General Thurman arrived on the spot to personally conduct the negotiations with Noriega and the Pope's Ambassador to secure the deposed Panamanian dictator's unconditional surrender. The 82nd Airborne Division's psychological warfare battalion also went into action, strategically placing loudspeakers blasting out Led Zeppelin's acid rock for the purpose of both demoralizing Noriega, who had rather refined musical tastes with a preference for Italian opera, and to drown out what was being discussed during the forty-eight hours of negotiations.

When Noriega finally stepped out of the front gates of the Papal Embassy anonymously dressed in jeans and a baseball cap, agents of the US Drug Enforcement Administration immediately handcuffed him, piled with him into a car, driving out under heavy military escort to the runway of Howard Air Force base and straight into the open back ramp of Major Skip Davenport's C-130 Talon. With his special operations plane now repaired from the damage it received at Río Hato, Davenport managed a very smooth four-engine take-off this time, flying directly on to Homestead Air Force base in southern Florida.

'We were worried when we flew the gap between Cuba and Mexico wondering what we would do if Castro sent MiGs after us,' recalls Davenport. Radio silence was strictly maintained. From his darkened suite, next to the elevated cockpit, the Electronic Counter Measures officer closely tracked Cuban radars as F-16 fighter jets vectored in by an AWAC flew out to provide escort. Noriega kept quiet most of the way until they were approaching Florida, when he asked to be allowed to change into his general's uniform, which was supposed to be in the trunk with the clothes and personal items which had been packed for him. But the DEA agents couldn't find it – or the keys for his handcuffs. Walking off the plane in his chains, the general had to face the glare of the TV cameras wearing the jeans and baseball cap in which he had fought his last stand.

18

SKYJACK

The GIGN had dealt with the deadly fanaticism of Islamic terrorists before. In 1979 a captain and two sub-officers had been dispatched on an urgent mission to Saudi Arabia to help dislodge several hundred armed Muslim extremists who had taken over the Sacred Mosque in Mecca. Thousands of pilgrims on their annual *hajj* to Mecca were being held hostage in a labyrinth of underground passages beneath the Mosque.

The very survival of the Saudi monarchy hung in the balance as Captain Paul Barril presented the Saudi National Guard with his plan to use paralysing gas to flush out the insurgents from an underground area of sixty square kilometres. The GIGN advisers trained hand-picked teams of Saudi soldiers in demolitions and room-clearing techniques to conduct the assault, which resulted in 2000 dead after an entire day of close quarter battle.

On 26 December 1994 the GIGN again faced a potentially disastrous encounter with Muslim terrorists when Captain Denis de Favier prepared his 'super Gendarmes' to storm an Air France passenger jet hijacked in Algeria by suicidal terrorists of that country's Armed Islamic Group (AIG).

News about the hijacking, which began at midday on 24 December, caught many members of the French counter-terrorist unit at home with their families for Christmas. However, the tall, angular-faced de Favier had not strayed far from Paris and his briefings, contingency planning and transport arrangements were methodically prepared by the time his men were assembling early that afternoon at their HQ outside Versailles.

Four terrorists of the AIG, the extremist faction of an Islamic fundamentalist coalition waging a bloody civil war against the military-backed government of Algeria, had managed to board an

Air France airbus disguised as airport security men. As the aircraft readied for take-off at 11.15 a.m., the terrorists had produced miniaturized AK-47s with collapsible stocks, crying 'Allahu Akbar' to announce that the airbus was under their control and all of the passengers and the crew were their prisoners.

With their blue strobe lights flashing, the GIGN vehicles raced out on to the forested highway leading out of Versailles, as 257 terrified hijack victims helplessly watched an Algerian policeman and a Vietnamese diplomat being dragged to the front of the aircraft. There were cries for mercy and then the crackle of automatic gunfire as the two men were shot in the back of the head and their bodies dumped on the runway. The terrorists were deadly serious about their demand to be flown to Paris.

The GIGN team, which by that evening had driven on to the military airfield of Neuilly, boarded an Air France airbus identical to the hijacked plane. Special arrangements had been made so that they could completely familiarize themselves with the aircraft's workings, and so plan and practise the counter-hijack operation, while flying south to their destination. But clearance to land at Algiers' Boumédienne airport had not been received as de Favier and his team took off shortly before 8 p.m.

Philippe Legorjus, the former GIGN chief who had trained de Favier and most of the other men now preparing for what was their first real-life storming of an aircraft, had also cut short his Christmas break. He was driving to his parents' home in Normandy with his wife and two children when his bleeper sounded. After dropping them off with hurried kisses, he drove without stopping to his reserved parking place beneath the corporate headquarters of Air France, off the Champs-Elysées in Paris.

As the chief security consultant for France's national airline, Legorjus was the vital member of the Air France crisis-management team which now came together. The company president was in touch with the French government, the chief pilot was finding out as much as possible about what was going on inside the plane and passing on the information to Legorjus, who was in constant radio contact with the GIGN team. He was also on the phone to the chief of Algiers' airport police and the sinister commander of the Algerian Army's elite commando force, the Ninjas, who ringed Boumédienne airport.

French Prime Minister Balladur flew to Paris from the Alpine resort of Chamonix, Interior Minister Pasqua arrived from the Côte d'Azur, and Foreign Minister Juppé chaired the government's swelling crisis team as a host of other government officials interrupted their holidays to assume their places. The decision was quickly reached to accede to the hijackers' demands by allowing the aircraft to land in Marseilles. But the Algerian authorities would not let it leave Algiers.

'The Ninjas wanted us to launch a full-frontal assault on the airbus,' explains Legorjus, 'using explosives to blast our way inside. It was going to be a disaster.' And probably similar to the fiasco a decade earlier when Egyptian commandos had tried to storm a plane hijacked to Malta, killing fifty hostages in the process.

Given the military nature of the Algerian government, certain key army commanders held sway over Cabinet ministers who would otherwise have been prepared to comply with the French government's wishes. Throughout Christmas Day heated exchanges by telephone went on between members of Balladur's Cabinet and their Algerian counterparts, and by evening the crisis seemed to be reaching a head. The terrorists had laid down an ultimatum. If the mobile stairway was not withdrawn and the airbus allowed to take off, they would kill a hostage every half an hour. At 9.30 p.m. they brought a French national up to the cockpit: a young cook employed at the French Embassy. They held a gun to his head. 'If you don't allow the plane to take off they are going to kill me,' he pleaded over the radio.

Pasqua called the Algerian Interior Minister and told him to withdraw the stairway. He in turn called the colonel commanding the black-hooded, machine-gun-toting Ninjas on the runway. The Cabinet minister gave the Ninja chief a flat order: 'Withdraw the passenger steps.' The reply was a swift 'Fuck you!' The Ninjas, locked into a savage three-year-old struggle against the Islamic guerrillas, could not agree to an order from civilians which could be considered as any type of concession. 'They didn't even want our help,' says Legorjus. Advice from two French military advisers at Boumédienne airport was being ignored and permission for the GIGN team to land continued to be refused. De Favier and his men were waiting on board their airbus at the nearby Spanish airport of Palma de Mallorca.

The Algerians tried to reassure the French government that the terrorists were bluffing. But shortly after 10 p.m. the French cook's corpse came tumbling down the mobile stairway. Discussions followed between Balladur and the Algerian Prime Minister, Mokdad Sifi, and between other key French negotiators and high-ranking

Algerian military commanders. Following direct orders from his Prime Minister and from army generals, the Ninja colonel disgustedly agreed to withdraw the stairway, allowing the plane to take off at 2 a.m. for Marseilles, where the French government had cleared it to land.

De Favier's team were already in place, having touched down just twenty minutes earlier on board their airbus at an airstrip adjacent to the main airport. GIGN snipers had taken up positions around the control tower. Paratroopers from another special surveillance unit, the Escadron Parachutiste de la Gendarmerie Nationale (EPGN), were camouflaged among the tall grass lining the runway, closely observing the hijacked plane, Flight 8969, taxiing to a halt at Marseilles-Marignane.

The negotiations were now entirely under the control of the GIGN. The Gendarmerie's chief prefect in Marseilles, Alain Gehim, served as the main interlocutor with the hijackers, speaking with them over the radio from the control tower. At every juncture during the discussions, de Favier would quietly pass him written instructions. The priorities were to find out as much as possible to give the counter-terrorism team the necessary time to prepare their assault and favourably position the airbus. They could conduct the assault as soon as they were ready – and the sooner the better. There were reasons. Flight 8969 could under no circumstances be allowed to fly to Paris or indeed leave Marseilles.

Intelligence officers at the French Embassy in Algiers had received an urgent call from a paid informant in the Islamic underground, saying that the AIG planned to blow up the airbus over Paris, crashing it into the middle of the city. The information appeared to have some corroboration from some of the sixty-three passengers who had been released in Algiers. Several said that they had heard the four gunmen, who seemed highly fanatical, talk repeatedly about flying to the eternal paradise and make many references to 'Allah's perpetual white light' – the Islamic vision of holy death. The possibility of a Lockerbie-type terrorist air disaster over the French capital was being taken very seriously.

Fearing that explosive charges had already been set on board the aircraft, the GIGN had to plan its assault very carefully, and it was now the government which was growing impatient, wanting action by mid-morning. But de Favier insisted that he needed to get right up close to the aircraft to plant listening devices to monitor the exact movements and sounds inside the plane before he could refine his assault plan. The intelligence being gained through microwave sound amplifiers trained on the aircraft was not proving sufficient.

Terrorist demands for twenty-seven tonnes of fuel to fly on to Paris were used as leverage by the GIGN to stall the hijackers. Gehim demanded in exchange the release of all the passengers. At one point the hijackers' leader, Abdul Yahia, screamed into the cockpit microphone, 'You want us to blow up everything right here! You have one and a half hours to let us take off to Paris.' The deadline he had set was 10.00 a.m.

It was a slip on Yahia's part, indicating that explosives were wired up inside the aircraft. The terrorists had maintained several hours of radio silence between their landing at 3.30 a.m. and the start of negotiations with Gehim at 6 a.m. The placing of two ten-stick packs of dynamite in the cockpit and beneath a row of passenger seats fused by one detonator had, apparently, been carried out during that time. The amount of fuel the hijackers were demanding was three times what was needed to fly to Paris. Its real purpose, the GIGN speculated, was to created an exploding fireball somewhere above the Eiffel Tower.

Gehim then spun out the negotiations still further, offering to provide the hijackers with food and to service the aircraft, cleaning it and emptying the blocked toilets, which were creating a terrible smell. They agreed. Disguised among the stewards and service crew who were allowed to the aircraft, GIGN operatives inserted minuscule eavesdropping devices. Tiny infrared closed-circuit cameras and cannon microphones were placed by windows along the exterior of the fuselage. It was now possible to monitor closely what was happening in the aircraft.

'We now reached the consensus to pass on to offensive action,' says de Favier. 'We had the necessary intelligence to plan our operation with exactitude. We knew that none of the entrances to the aircraft were booby-trapped or obstructed and that two terrorists were inside the cockpit at all times.' It was decided, however, to try to press for the release of more hostages before an assault which would inevitably entail casualties. There was no doubt that the terrorists were determined to resist and were well armed. Each had a 7.62mm Kalashnikov sub-machine-gun and hand-grenades. Most menacingly, there were also the explosives.

Shortly after midday an elderly couple walked off the airbus, leaving over 150 fellow-passengers on board. Yahia insisted that it was the last concession he would make before reaching Paris. The GIGN assault force began positioning themselves around the runway. After dividing his men into four teams, de Favier moved them into predetermined locations hidden from the view of the aircraft. Using careful calculations, he had planned the exact angles from which to rush the plane's three main entry points on mobile stairways. The GIGN commander was confident that access into the fuselage could be gained by simple manipulation of the hatch doors' exterior mechanisms.

The released prisoners told the hostage negotiators that the hijackers were growing increasingly impatient and irritable, reading out passages from the Koran in frantic tones over the aircraft's speaker system.

By 4 p.m. the GIGN were in position at the bottom of their respective mobile stairways and fully prepared. After putting their Kevlar helmets with yellowish Plexiglas visors over their heads, already hooded with ski masks, and strapping bulletproof vests over fire-retardant suits, they checked and rechecked their .357 magnum revolvers, and their 9mm MP-5 sub-machine-guns and automatic pistols. Snipers lying low at the top of the control tower trained their rifles' telescopic sights on the cockpit, where two of the terrorists guarded the pilot and flight crew. The assault could be launched at any moment, but with the patient assurance instilled by experience and cool professionalism, Captain de Favier preferred to wait a little longer. He wanted to wait until twilight.

'The biggest problem until now had been to stay concentrated amidst the inaction and uncertainty. We could anticipate the danger although we did not really imagine what awaited us,' recalls Alain, who had barely completed two years in the GIGN and was now among the lead squad of Gendarmes who would be entering the hijacked plane's cockpit.

Just before 5 p.m. de Favier's instructions to be ready to move at a moment's notice came over the radio sets inside the men's helmets. Magnums were unholstered and those men with MP-5s pulled back their round cocking levers. Suddenly the searing noise of accelerating jet engines came from the airbus parked some 400 yards away, and it started to move. Yahia's shrill voice then came screaming over the control tower's amplified system. If the plane was not refuelled for take-off in fifteen minutes, he threatened, 'I will take action!'

Prefect Gehim reassured him that the refuelling trucks were on the way as the aircraft rolled slowly towards the control tower. He played for time while the GIGN assault teams repositioned themselves, adjusting their angles of approach. The plane got to within thirty yards of the control tower and at 5.08 p.m. a gun barrel protruded from the cockpit's window. Shots cracked out and Gehim ducked, together with the others in the tower, as 7.62mm rounds shattered the bay window before them. The snipers held their fire. Gehim strained to keep his voice calm over the next few minutes, trying to sustain his dialogue with Yahia as the GIGN adjusted their plan.

'Zero,' said de Favier finally. His signal for action rushed through his men's headsets like an electrical current, triggering, as if by remote control, a series of carefully choreographed split-second movements. The trailer engines of three mobile stairways started simultaneously. The groups of eight Gendarmes positioned on each one climbed the steps in double file as the stairways drew ever nearer to the plane. A sniper on the control tower repeatedly squeezed the trigger of his .50 rifle, its silencer muffling the six shots that smashed through the cockpit's right-hand windscreen. 'He had to be careful,' it had been made clear. The heavy rounds were fired high. 'It was very difficult to determine who was terrorist and who was air crew inside the cramped cabin, where everyone ducked for cover as the half-inch bullets crashed inside, distracting the terrorists as the stairway came up to the right-hand door of the aircraft.'

One Gendarme leaped to grab the hatch. Sliding with the door to the side to make sure it stayed open, he let his feet fly off the stairway. A GIGN stun grenade was thrown inside and flashed through the first-class section as a shooter pair, one armed with a magnum and the other with an MP-5, rushed through the narrow opening. They caught two shocked hijackers along the hallway, making for the cockpit. One shot from the magnum instantly blew away the nearest terrorist, drilling a .357 bullet diagonally through his forehead. But in the next half a second a barrage of automatic fire from Yahia and another hijacker barricaded with the flight crew, hit the Gendarmes like an avalanche of steel.

Squeezing a burst from his MP-5 after the other terrorist leaping into the cockpit, Sub-officer Thierry pursued him through the open door, resting his sub-machine-gun against his shoulder to carefully train its sight and avoid hitting hostages. But he instantly received seven Kalashnikov rounds. Three 7.62 bullets perforated his exposed right arm, shooting off three fingers curled around the MP-5's hand grip, which was also shot to pieces. Other bullets hit his shoulder and chest, shredding the black vinyl fabric and denting the steel ceramic plates of his bulletproof vest. Another round ricocheted off the Gendarme's Kevlar helmet as he reeled backwards, collapsing on to the carpeted hallway. He could only think of protecting himself from the continuing barrage of bullets and heard someone shout 'Grenade!' as an explosion ripped through the cabin, wounding the other Gendarmes rushing in behind him. Thierry felt the pain of shrapnel peppering his right leg before losing consciousness.

Eric, who had shot the first terrorist, had the magnum blown out of his hand by a bullet hitting its long steel barrel just as he fired a second round. In the next split second he too was lying on the floor, his face bleeding from the helmet's splintered visor – proofed against 9mm ammunition but unable to withstand the high-velocity 7.62 rounds. Feeling other wounds along his shoulder and arm, he could still only think of recovering his weapon, which he found lying by a corner, its handle blown off by the grenade explosion.

The fire from the AK-47s was incessant. 'I'll never forget their characteristic clapping noise as the bullets flew around me. It was impressive,' says Olivier, who entered the cabin amid the smoke and shrapnel of the fragmentation grenade. Jolted by the impact of several rounds on his bulletproof vest, he was soon partially immobilized by trauma to his spine and a bullet in his left hip.

'Six of my men fell wounded around me as we came inside,' recalls de Favier, who personally led the first assault wave and was the only one who had not been hit as Yahia's hate-filled face peered through the cockpit's cracked door as he fired his sub-machine-gun. De Favier instantly aimed his magnum, but a bullet from an AK-47 hit the side of his helmet, tearing off the visor and forcing him to take cover. Hundreds of rounds, many coming straight through the door and dividing walls of the cockpit, were hosing the Gendarmes. 'The terrorists were highly trained,' says de Favier, 'and very quick.' They hardly took a second between releasing spent ammunition clips and inserting fresh ones into their curved magazines. They knew exactly how to fire, aiming for heads and the upper chest. 'Their deadly spray didn't let up for an instant.'

The Gendarme who had opened the right-hand door of the airbus had remained hanging in the air from the handle as the seven others rushed inside. He then let go. In the twenty-foot fall on to the concrete runway, he used his parachute training to relax his legs, and landed uninjured. As he reached the cabin door amid a hail of gunfire and with bodies lying everywhere, he went into a low crouch, aiming his lightweight Austrian-made 9mm automatic pistol and squeezing off several rounds into the cockpit door. Suddenly a 7.62mm bullet penetrated squarely through his pistol's stock, going beneath the barrel, driving into the chamber and coming out on the trigger, which disintegrated into tiny pieces as the weapon flew out of his hand. The Gendarme was left with a broken index finger as another bullet smashed through a spare ammunition clip in the chest pouch of his bulletproof vest.

The front left-hand door of the aircraft now slid open as the eight Gendarmes in the second assault wave stormed into the smoking inferno, which stank of blood and cordite. A terrorist shooting from the cockpit's side window sprayed them with a burst of sub-machine-gun fire before they were even inside. Several rounds hit the side of the sliding door as it moved towards the cockpit. Aiming his MP-5, the lead gunner instantly followed de Favier's orders, firing at point-blank range at the left section of the cockpit's wall, through which a lot of gunfire was coming. But AK-47 rounds fired back at floor level hit the Gendarme's lower legs within point-blank range of the partition. The officer fell and had to be pulled away under covering fire by other Gendarmes, who began evacuating their most severely wounded comrades.

GIGN men were simultaneously moving through the rear of the aircraft from the tail doors. Having opened emergency exits, they evacuated the 159 passengers, who came sliding out on escape chutes as the gun battle raged on at the front of the aircraft. Stray bullets flew into the main seating section but no passengers were wounded by gunfire and only thirteen were treated for light injuries, mostly cuts and bruises after the full evacuation was completed in twenty minutes.

The moment he saw blood bursting out of the terrorist closest to him as the intensifying stream of GIGN fire filled the cockpit, the plane's navigator calculated that the surviving gunmen would be too distracted to notice as he dashed to the open side window and jumped out of the plane. In doing so he was severely injured, but was quickly picked up by ambulance and rushed off for treatment for a broken leg. The pilot and co-pilot remaining in the cockpit hugged the floor beneath their seats, trying to keep their heads down as the body of a dead terrorist fell on top of them. Although wounded, Yahia kept inserting clips into his Kalashnikov, frantically spraying fire through the bullet-riddled partition.

The ferocious battle continued for ten minutes as two more Gendarmes from the second wave were wounded. Unable to see their targets, they followed de Favier's directions, always aiming their shots at the sections of the partition through which the gunfire was coming. Finally they heard the pilot's voice from inside, screaming, 'Stop shooting!' Bursting into the cabin, the GIGN found the bullet-riddled bodies of three dead terrorists lying on top of both pilots, who were, remarkably, unscathed amid the blood, wreckage and piles of spent shell cases. They could see the package of dynamite tucked beneath the pilot's seat. It was fused to a detonator but the terrorists had not been given a moment's chance to operate the fuse box mechanisms.

The GIGN's action at Marseilles has been hailed as the most dramatic counter-terrorist operation since the SAS debut at Princes Gate. In the view of retired GIGN commander Philippe Legorjus, 'It was the most important intervention in the history of counter-terrorism. It was wonderful to see all the men whom I had trained and led perform so well under that kind of pressure.'

There is little doubt that the wild Russian roulette of international terrorism had dealt the GIGN a loaded bullet. Nevertheless, the cost to the special operations unit was relatively low, considering the potentially disastrous situation they had faced: four suicidal, well-armed and expertly trained terrorists on an aircraft with around 160 hostages wired to explode with dynamite. Only Sub-officer Thierry remained hospitalized three months later, permanently disabled with one finger missing but feeling fortunate that he wasn't missing anything else. Three more men who underwent intensive care, Eric, Alain and Olivier, were released from hospital shortly after a month's stay. The six other wounded Gendarmes needed no more than superficial surgery.

The GIGN operation on board Flight 8969 was classically unique in that it succeeded almost entirely in close quarter battle, the very essence of special forces counter-terrorist training, but with the added complication in this case of hardly being able to see the enemy. Unlike previous episodes, such as Princes Gate or Djibouti, stand-off sniping played a very limited role in Marseilles. Here the assault team overcame the entire terrorist group, with hostages in their midst, inside a severely enclosed space, without causing the death of a single innocent victim.

A month later de Favier was still feeling the rush of adrenalin-fuelled terror and excitement, never quite forgetting the taste of blood as he had levelled his magnum at Yahia while his men fell around him and a Kalashnikov bullet tore into his helmet. The telephone rings and he answers. After a brief conversation the young super commando, attired in a tweed jacket and flowery tie, hangs up. 'That was the US Embassy,' he says. 'They want to arrange an international meeting of special intervention forces here in Paris.' He looks forward to the exchange with his counterparts from the SAS, Delta and the SEALs. 'I'm not a hero,' he adds. 'I just like my job. Next time it could just as easily be the turn of any of the others.'

INDEX